James John Howard Gregory

Pamphlets on Vegetable Gardening

James John Howard Gregory

Pamphlets on Vegetable Gardening

ISBN/EAN: 9783337375294

Printed in Europe, USA, Canada, Australia, Japan

Cover: Foto ©berggeist007 / pixelio.de

More available books at **www.hansebooks.com**

CARROTS,

MANGOLD WURTZELS

AND

SUGAR BEETS.

HOW TO RAISE THEM, HOW TO KEEP THEM
AND HOW TO FEED THEM.

By JAMES J. H. GREGORY,

AUTHOR OF "ONION RAISING," "CABBAGE RAISING," ETC.

MARBLEHEAD, MASS.
MESSENGER STEAM PRINTING HOUSE.
1882.

CONTENTS.

	Page.
The Argument for the Raising of roots,	3
THE CARROT	5
The Location and Soil	6
The Manure and its Application	7
Preparing the Bed	12
When to Plant	13
The Seed and the Planting of it	14
Quantity to the Acre	15
Varieties, and What Kinds to Grow	17
Early Very Short Scarlet	19
Early Short Scarlet Horn	19
Short Horn	19
Danvers Carrot	19
Long Orange, or Long Surry	20
Altringham	21
Large White Belgian	21
The Cultivation, and the Implements needed	22
Gathering and Storing the Crop	24
Raising Carrots with Onions	27
Marketing and Feeding	28
THE MANGOLD WURTZELS	30
Varieties	32

The Long Varieties	32
The Round Varieties	33
The Ovoid Varieties	33
What Kinds to Grow	33
The Soil and its Preparation	36
The Manure and its Application	37
Salt as an Auxiliary Manure	45
Planting the Seed and Tending the Crop	46
Gathering and Storing the Crop	49
Feeding the Crop	52
The Cost of the Crop	57

CARROTS.

THE ARGUMENT FOR THE RAISING OF ROOTS.

The fact that the most progressive and successful farmers in the dairy districts, where the prices received for the products of the dairy stimulates to the highest enterprise, are raisers of roots, (by which I now more especially refer to Carrots or Mangold Wurtzel) in about the same degree as they are pecuniarily successful, is in itself a great practical argument for root culture.

In nutritious value roots compare with hay in about the average proportion of one to three. If now we consider that thirty-four tons of Swedes, nearly forty tons of Carrots and seventy-four tons of Mangold roots have been raised in Massachusetts, to the acre, and that to each of these crops should be added at least 15 per cent. for the fodder value of the yield of leaves, which were not included in these estimates, we have a demonstration of how immensely more is the nourishment that can be obtained from an acre of roots than from an acre in hay. Such an immense increase in the nourishing products of the farm, if fed on the premises as it should be, unless the farmer is so located that he can buy manure cheaper than he can make it, means a great increase in the manure products, and consequently a great increase in the crops,—so that it has been wisely said, root culture lies at the basis of good husbandry.

Carrots and Mangolds are subject to but few diseases. In discussing the nutritious value, chemists differ somewhat, according as they measure this by the nitrogen they contain, their per cent. of dry matter or sugar, but they agree in ranking them much superior to the early varieties of turnip and somewhat superior to the Ruta Baga or Swede class, particularly when fed to full grown cattle. Prof. Johnson ranks Carrots with Cabbage when fed to oxen, for nourishment, and experiments appear to have proved that when equal measures of each are fed, Mangolds will give a greater increase of milk than potatoes, by about a third. For some reason not fully understood, (perhaps the depth they penetrate the soil has something to do with it ;) Onions will do better after Carrots than after any other crop, the yield being larger, the bulb handsomer, while the crop will bottom down earlier and better. Unlike Turnips or Swedes, with high manuring the crop can be profitably grown for years on the same piece of land. Swine prefer Mangolds to any root except the parsnip, and both in this country and in England store hogs, weighing from 125 lbs. and upwards have been carried through the winter in fine condition, when fed wholly on raw Sugar Beets or Mangolds. Chemists rank Carrots, when compared with oats, with reference to their fat and flesh forming qualities as 1 to 5.

Not only have roots a value in themselves as food, but they have a special office. taking to a large degree the place of grass and preventing the constipation that dry feed sometimes causes. While practice proves that they should not be relied upon to entirely supersede hay or grain, still they increase the value of either of these to a large degree ; and for slow working stock they may be fed with profit in place of from a third to half the grain usually given. Carrots add not only to the richness of the color, but also to the quality of the milk, while the flavor of the butter made from such milk is improved. Carrots fed in moderate quantities to horses give additional gloss to their hairy coats, and have not only a medicinal value when given to such as have been over-grained, but aid them in digesting grain, as may be seen in the dung of horses fed on oats with Carrots, and that of those fed on oats without Carrots. When cooked they are sometimes fed to poultry, and either cooked or raw to swine. In the family economy they have their place, particularly when young and fresh, while in Europe they enter largely into the composition of the well-known vegetable soups of the French.

THE CARROT.

"The Carrot," (*Daucus Carota*) says Burr in his "Field and Garden Vegetables of America," a book worthy a place in every farmer's library.—"in its cultivated state is a half-hardy biennial. It is indigenous to some parts of Great Britian, generally growing in chalky or sandy soil, and to some extent has become naturalized in this country; being found in gravelly pastures and mowing fields, and occasionally by roadsides, in loose places, where the surface has been disturbed or removed. In its native state the root is small, slender and fibrous or woody, of no value, and even of questionable properties as an article of food."

The average result of several analyses of the Carrot as given by Dr. Voelcker, is as follows:—

Water, - - - - - - 87.0
Albuminous Compounds, - - - - .7
Fat, - - . - - - .1
Pectine, - - - - - 1.6

Cellular Fiber,   3.5
Sugar,   6.5
Ash,   .9

THE LOCATION AND SOIL.

It is important in selecting a location for the Carrot bed that the land should be nearly level, as otherwise the seed will be liable to wash out after heavy showers, and the plants while young be either washed out or covered with soil and killed. The land should be clear of all large rocks, and as far as possible of all stones up to the size of a hen's egg. The presence of large rocks 'in place,' as the geologists say, would interfere with the continuity of the rows, while the loose stones are not only always in the way while raking and planting the bed, but are also in the way of the slide or wheel hoes which are apt to knock them against the young plants to their injury. It is important that the piece of ground selected for a crop that will require so much manure and labor should have every advantage possible in its favor; it should not only be level and comparatively free from stones, but if possible should have been previously under high cultivation, that it may come to Carrots when in high condition.

The best soil, particularly for the Long Orange variety, is a loam mellow to the depth of two feet or more. On such soil the Carrot will perfect itself, growing straight and altogether beautiful to look upon, as they stretch from side to side of the bushel boxes. On some market gardens near critical markets, farmers find it for their interest to ascertain by actual experiment on what part of their grounds the root will grow longest and straightest, and when such plot is found make it a permanent bed. If the soil does not naturally grow a long carrot and they are desired, the end may be attained by trenching deep and adding sand. The difference in the shape of the Long Orange, when grown on a deep mellow loam, and on a heavy soil with a compact sub-soil, is so remarkable that it would be almost impossible to make an inexperienced person believe each lot was from the same seed,—those grown on the heavy soil, resting on a compact sub-soil, oftentimes so closely resembling the Intermediate varieties as not to be distinguished from them. Though the course is not on the whole to be advised, yet Carrots can be raised on freshly turned sod. Such land will be very free from weeds, and by making good use of the wheel harrow, and applying manure in a very fine state, should the season be a moist one, fair crops may be raised. Reclaimed meadows in a good state of cultivation, which are well-drained to the depth of thirty inches, will oftentimes grow crops, large in bulk, but the individual roots are oftentimes inclined to "sprangle," and unless such meadows have been well drained, and liberally covered with sand or gravelly loam, they are apt to be spongy and inferior. When grown on land inclining to clay, they are apt to be small and woody in structure ; still, such land, if made friable by good underdraining and the application of sand may be made fair carrot ground.

THE MANURE AND ITS APPLICATION.

All root crops delight in most liberal manuring and the highest of cultivation. Carrots are no exception to this rule. With every crop, other conditions being equal, *it is the last half of the manure gives the profits;* and the more costly the cultivation required the more important it is that this golden fact be borne in mind. Though chemical analysis shows difference in the composition of all roots, and that there is therefore an office for special manures, yet their general composition is so nearly alike, and animal manures, most of which contain in greater or less proportion, all the elements required, are so difficult to handle in just the proportions that would be required from the chemical standpoint, particularly when we consider that soils on which root crops are grown are usually rich in manures, varying in their chemical constituents, left over from former crops ;—for this reason I treat of manure by the cord and with reference to its comparative strength, bulk for bulk, rather than of its chemical elements.

Eight cords of good stable manure, nine cords of a compost made of one part night soil to two parts muck or loam ; twelve cords of a compost made of one-third fish waste, by which I mean the heads and back-bones of the fisheries, and two-thirds soil; eight cords of muscle mud; six or eight cords of rotten kelp—either of these applied to an acre of land in good condition by previous high cultivation would be sufficient for a good crop of carrots. Other manures might be mentioned, but these will serve as a pretty good measure of value for any kind accessible to farmers in general. To produce a very large crop such as one would like to be able to point to when premium crops are called for, add from one quarter to one-half to the above quantities. The condition of the manure is a matter of importance ; the stable manure should be good; not half bedding, not burnt, neither too coarse nor too new ; the night soil should have been well mixed with the soil in the compost heap, and have been pitched over twice with sufficient intervals between to allow it to develop some heat. The fish waste should be well decomposed, so well that all but the bones should have disappeared, and if these be very dark and brittle so much the better. The muscle mud should be rich in dead muscles. In all farming it is important that the manures applied should be in a fine condition mechanically, and particularly is this true of root crops. For the roots of all plants can take up only such parts of the manure as are dissolved in water, and the finer the manure is the more readily can water penetrate it.

A man who is unfortunately short of manures can materially increase the capacity of what he has by working it over until it is very fine.

When short of a supply of animal manure, guano and good phosphates, where the soil is already in good condition can be used with success, provided the season does not prove to be too dry a one. From eight hundred to a thousand pounds of Peruvian guano and from ten to fifteen hundred pounds of the best phosphates should be used. The famous fertilizer formulas of Prof. Stockbridge have generally done so well I should be willing to try them on an acre of Carrots, were I short of other manures.

There is another matter concerning our manures which requires attention; if they are too fresh or crude they will be apt, if applied to our long growing varieties, to drive the growth too much into the top of the Carrot, to the loss of the root, giving us tops to our knees with roots about the size of a hoe handle. It is important therefore, when used liberally, that they should be somewhat decomposed—that the mixtures should be *composts*, as far as the time will allow, and not mere mixtures. To the shorter varieties the crude manure may be applied with a degree of safety. Here let me note a fact that I think is of general application in farming, viz. :—that a style of manuring that will drive tall growing varieties of vegetable nearly all to tops or vine, with dwarf varieties of the same kind will work admirably. The Pea is a very good illustration; to get a good crop of Dwarf Tom Thumb, manure liberally, but the same quantity applied to the taller sorts would drive them excessively into vine at the expense of the crop.

Don't make your compost heap on the ground where the crop is to grow, for the result will be no crop where the heap stands. For the same reason it is bad policy to cart out any strong manure to stand on the land in heaps, no matter how small, over winter. There will be nothing lost by spreading the manure over the surface before the ground is frozen. In getting it into the soil, *keep it as near the surface as possible* without its interfering with the planting of the seed, bearing in mind the nitrogen, that element in manures, about the loss of which by evaporation there is much uncalled for anxiety tends to work down into the soil. If the manure is coarse it may be applied to the surface in the Fall and be deeply ploughed in, and in the Spring again brought to the surface by ploughing equally deep, having meanwhile received the benefits of frost and moisture.

In applying guano or the phosphates, keep them near the surface, scattering them broadcast and raking or harrowing in. It is best not to apply either of these all at once,—particularly is this true of guano. Apply about half at the time of sowing, and the remainder when the crop is about one-third grown—following it with the slide hoe, which will tend to work it in just under the surface. In applying guano and all similar fine manures in the Spring time, it is well to do so early in the day, as winds are apt to rise as the day advances, which seriously interfere with the economical application and even distribution. Both phosphates and guano tend to hasten the maturity of the crops to which they are applied. There is one condition that has a very important bearing on the cost of Carrots and all roots, viz. :—that both the ground and manure should be as free from all weed seed as possible. For this reason ground recently from the sod, the third year, provided it has been kept under a high state of cultivation, and such manures which from their very nature must be comparatively free from the seed of weeds, such as fish composts, night soil, or barn manure a year old, are to be preferred.

Dr. Voelcker gives the result of 10 analyses of the ashes of the root and 2 of the ashes of the leaves of the Carrot, and from these deduces the following as the number of pounds of mineral matter taken from an acre of land, by 10 tons of roots and 4 tons of tops.

Potash,	Soda,	Lime,	Phosphoric Acid,
116 lbs	86 lbs.	101 lbs.	31 lbs.
	Sulphuric Acid,		Chlorine,
	34 lbs.		31 lbs.

To those who desire to experiment with mineral manures this table will be interesting as showing the kinds and proportion of each needed. The potash is found in unleeched ashes, at the rate of 4 or 5 pounds to the bushel; or in the German Potash salts; the soda and chlorine in common salt, (chloride of sodium); lime in the common lime of the mason, the Phosphoric acid in the phosphates offered in the markets, and the Sulphuric acid in that directly or in common finely ground plaster, known by chemists as Sulphate of Lime.

I shall have occasion to present some very valuable suggestions of the learned Professor, under the head of "The Manure" in my article on Mangolds, to which they more especially apply.

The greatest single item in the cost of any crop is the manure, but this is an exceedingly varying element. Farmers near cities, and particularly if they also reside near the sea-coast, as an off-set for the greater cost of farming-land and expenses of living, have the advantages of a city market and special facilities for collecting manures, at a cost to them, much below the standard value of stable manure. Night soil to almost an unlimited extent, can be obtained for the cost of collecting it, while the waste material of the fisheries, Kelp, Rock Weed, Muscle Mud, Glue Waste, Sugar House Waste, and the products of the distilleries, these and other rich fertilizers can be procured at so low a figure, in proportion to their value, that root crops can be raised considerably cheaper than in farming districts not so favored. Many a man can be found in these favored districts who thinks he is making a good business at farming, yet could he but sell the manure he gathers so cheaply, at its market value, barn manure being the standard, he would make money by doing so and folding his arms the rest of the year. The fact is he is really losing money at farming; but through his crops he is selling what cost him but a trifle, at a price, indeed, below its real value, but still so far in advance of cost as to leave a profit. Such a man does wisely in the course he pursues though he makes a mistake in the debtor and creditor side of the account, for it is most decidedly wiser to be at work than idle, though the result makes no difference in the dollars in a man's pocket.

PREPARING THE BED.

The great object here should be to get the soil thoroughly fine that the small, thread-like fibers, and the roots themselves, may waste the least possible vital power in permeating the earth in search of food, or in pushing downwards. The vitality wasted in this way is just so much taken from growth, and may make the sole difference between a good crop and a poor one. If it is necessary that the first ploughing should be a very deep one, better apply the manure, (as previously stated, the finer mechanical condition this is in the better) afterwards. Should the manure be to any degree coarse after spreading, run the brush or wheel harrow over it, one or both. This will also break up the clods and fine up the soil and incorporate the manure with it. If still at all lumpy, follow with a plank drag. Next plow shallow a few furrows, and have men, with wooden-toothed hand rakes, rake at right angles, pulling all coarse stones and lumps of earth and manure into the last furrow made. In brief, proceed to make as fine a seed bed as for onions.

If any one, depending on the apparent fineness of the surface, concludes to dispense with the final raking and let the work of the brush harrow answer, he will be apt to repent it before the season closes; should he try it let him be sure to double the quantity of seed planted in that portion of the land so treated. If the bed has its first ploughing early in the season, much of the weed seed will germinate before planting time and an occasional use of the cultivator will destroy many of the pests.

WHEN TO PLANT.

Some of our best farmers advocate planting about the middle of May, others equally successful in root culture claim that the middle of June is the best time. There are arguments for both early and late planting. In New England we usually have the weather sufficiently moist towards the close of May to insure the germination of the seed and protect the plants when they break ground, from "sun-scald." Those planted as late as the middle of June are more liable to be so affected by the dry weather usual at that period as not to vegetate as well; and should the heat be very great just after they push through the ground, sometimes in a single day nearly the entire crop will disappear by "sun-scald." But on the other hand, by planting late we about get rid of one weeding, assuming that the ground is stirred by the cultivator occasionally, up to the time of planting. Again, this brings the crop in full vigor in October, the month of all others most favorable for the growth of the root, and the Carrots being dug while the tops are in fair growing condition, keep better than when dug fully ripe. The argument for late planting holds especially good for the Short Horn varieties, as these require a shorter time to mature than the long kinds. If the crop is planted too early, sometimes the roots having matured, will attempt to push seed shoots; when this is so they will be found woody in their structure, with numberless thread-like roots while their quality and keeping properties are greatly injured. This crop on rich land is sometimes planted as late as the first week in July, and with great success, should the Fall prove exceptionably mild, yet, as a rule, I would not recommend planting later than the middle of June. If it so happens, from press of work, or the dry weather, the farmer has to plant later than this, then by all means let him confine himself to the earlier varieties.

THE SEED AND THE PLANTING OF IT.

To grow seed, medium-sized roots should be selected, that are well-grown, straight and symmetrical, of a rich, dark orange color, with a small, compact top. Plant in rows three and a half feet apart and fifteen inches in the row, the crowns being on a level with the surface. If the roots are long they may be laid slanting in the furrows. The best seed will be from the two first cuttings, which will come from the center of the main stock and of each side shoot.

The seed grows with a covering of small, short, stiff hairs, which makes them adhere together; these must be very thoroughly removed before the seed can be relied upon to flow freely from the machine. Much of foreign grown seed reaches this country not properly cleaned. To remove this furze, either thrash the seed with the flail very thoroughly, when the weather is quite cold and dry, or warm the seed slightly and rub it with the hand against the wires of a sieve, of a right degree of fineness to let the hairs fall through. Either winnow, or sink in water, to remove all impurities. If sunk, be careful to dry the seed at a very moderate temperature; rubbing with plaster, charcoal or earth dust will absorb what moisture may remain when nearly dry. As Carrot seed vegetates somewhat slowly and the plants are quite small when they first appear, weeds are apt to get the start of them before the rows can be seen with sufficient distinctness to make it safe to use the slide hoe. For this reason many farmers practice soaking the seed in water and keeping it at a temperature that will nearly develop the sprout, before planting. This may be done by soaking the seed from 36 to 48 hours in milk warm, rather strong manure water, then removing it to where the air is of about the same temperature. Stir it slightly for a few days, and finally dry it sufficiently to drop freely from the machine by adding plaster, charcoal or dust. Camphor has a wonderful effect in stimulating the vitality of seed, and the addition of a small quantity of it to the manure water would doubtless be of advantage. This process should not be carried so far as to develop the sprout. Should the surface of the ground be very dry when the seed is sown, this soaking process may be fatal, for if the germ is once started it will not live in a dormant state; it must either grow or die: whereas, seed that have not been soaked will vegetate after rains wet the dry surface. Be sure that the seed planter has a good roller attached to it, and not a mere coverer, as this will help confine the moisture and thus materially aid in developing the seed.

QUANTITY TO THE ACRE.

Tables vary greatly, some advising as high as four pounds to the acre. If the design is to raise small-sized roots for early marketing, possibly this might not be an excess of seed, but to advise so heavy seeding for ordinary field crops, means that much of the seed is poor trash, probably old and worthless, and put in as a make-weight.

Some years ago a party wrote me, offering a variety of garden seed at a very low figure, and stated that it was of his own raising. As it was a kind that I was in the habit of raising, I had the curiosity to write and ask how he could afford to raise it at such a price. He replied that it was of his own growing, but so old as to be good for nothing, and therefore he sold it to seedsmen at a very low figure, to mix with their good seed to *help make weight!* When four pounds of Carrot seed are advised to the acre, for a field crop, I think that some of this kind of seed must somehow have got into the mixture. With everything favoring, and the farmer by experience having his seed sower under perfect control, rather less than a pound of seed will be sufficient for an acre. The great object to aim at is, while having the plants thick enough, not to have much of any thinning to do, as it costs about as much to thin a crop as it does to weed it, with the drawback that the plants left in the ground are more or less started, and so put back by the thinning. As a general rule I would advise one and one-half pounds of seed to the acre, and this the farmer can reduce in proportion as he is favored by circumstances and advances in experience.

Twelve inches is a sufficient distance between the rows of the two small, early varieties, and fifteen between the rows of all other sorts. With the greatest of care the seed will not come up with mathematical precision. Some advocate leaving a plant to about every inch of row; others, to thin to four inches apart. Carrots are somewhat like Onions in their aptitude to grow to a good size when crowded, pushing out either side of the rows, and at times crops will give great bulk when the plants are nearer each other than four inches, still, as a rule I advise thinning to near this distance, leaving them thicker near vacant places.

VARIETIES, AND WHAT KINDS TO GROW.

Foreign catalogues give lists of about two dozen varieties, which differ in earliness, size, color, form, termination of root, characteristic of growing entirely under or partly above ground, and in the size of the core or heart. In foreign catalogues, what we call "Orange," are known as "Red" Carrots. From a test of these varieties I have thus far found nothing worthy of being added to the kinds already grown to a greater or less extent in the United States. The yellow-fleshed sorts are repudiated in New England by general consent; yet the Yellow Belgian, on a limited trial has proved with me, to be an exceptionably good keeper. The Purple or Blood-Red is of a deep purple color, a poor cropper and by no means attractive to the eye. The remaining varieties may be classed as follows:—Early, middling early and late. The first class is made up of the Early Very Short Scarlet, and the Early Scarlet Horn. The second class, of all the half-long or short horn varieties, and the third, of the long varieties, such as Long Orange, Belgian and Altringham sorts.

In addition to about one-half of these foreign varieties, cultivated more or less generally in this country, there are several kinds catalogued by seedsmen, all of which are but improved strains made by careful selections, through a series of years, from what was originally imported stock. These strains usually bear the name of some person. A brief discussion of the more valuable varieties will now be in order. Here I will lay down three general facts, viz.:—1st, that of the various orange colored varieties, the shorter growing kinds are, as a rule, the darker colored and sweeter flavored. 2d, that the proportion of dark, orange-colored roots in any crop, while it will depend largely on the care that has been used in the selection of seed stock for a series of years, does not turn wholly on this, but soil, season or manure, one or all have some influence in this direction. 3d, that the fact that more or less of the Carrots tend to push seed shoots the first year, while with the long varieties it may prove that the seed has been allowed to mix with the wild varieties, yet the probability, (marked cases excepted,) is decidedly the other way; while with the short horn varieties this tendency to push seed shoots the first season, so as to make something of a show when an acre is glanced over, is quite a common characteristic with seed of the very purest strain.

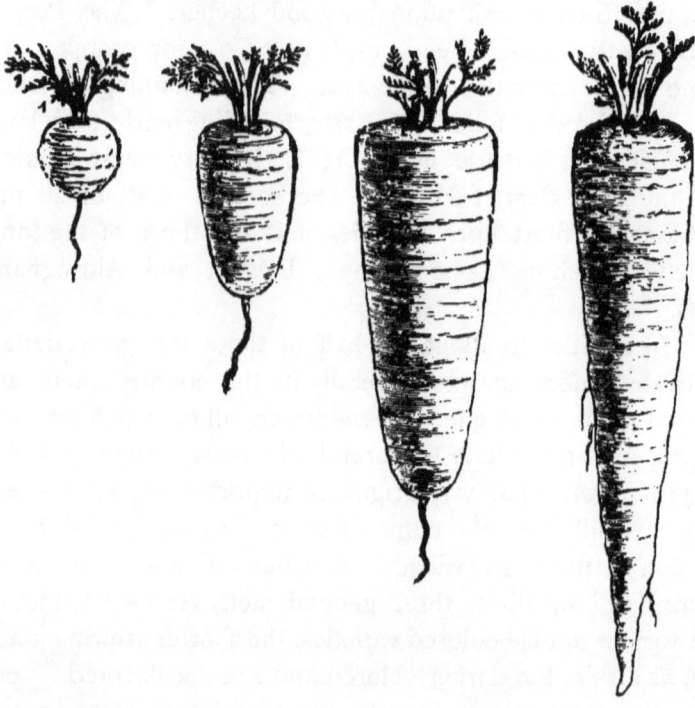

EARLY VERY SHORT SCARLET.   EARLY SHORT SCARLET.   SHORT HORN.   LONG ORANGE.

Early Very Short Scarlet. (see engraving.)

Early Short Scarlet Horn. (see engraving.) These two varieties are the shortest grown and are raised at times in forcing beds, for an early market, the former very generally so. They are of a very rich orange color, fine-grained, sweet, and of excellent flavor, heading the list for quality. Their rich color makes them valuable above all other kinds for coloring butter. Though quite short, yet the Early Short Scarlet Horn can be grown to yield a great bulk of roots, from the fact that from the smallness of their tops the roots can be grown very thick, two or three abreast all along the rows. When the small, handy size of this variety is considered in connection with the superior quality, it stands foremost as a table Carrot, and I therefore recommend it in preference to all others for family use.

Short Horn. (See engraving.) This variety, intermediate between the Early Forcing and Long Orange, with but slight variations in form, is shown under various names, as Intermediate, Nantes, Half Long, James' Improved, Stump-Rooted, &c. It is characterized by a darker color than the average of the Long Orange, finer grain, and a sweeter and richer flavor. In part from the more solid structure of the Carrot, and in part from its better stowage, thirty-six measured bushels of this variety make a ton, while of the larger varieties forty bushels are required. The best strain of this variety is doubtless the kind known as the "Danvers" Carrot.

Danvers Carrot. In the town of Danvers, Mass., the raising of Carrots on an extensive scale, has for years been quite a business—the farmers finding a large market in the neighboring cities of Salem, Lynn and Boston. After years of experimenting they settled upon a variety which originated among them (as did the Danvers Onion) known in their locality as the "Danvers Carrot." It is in form about midway between the Long Orange and Short Horn class, growing very generally with a stump root. The great problem in Carrot growing is to get the greatest bulk with the smallest length of root, and this is what the Danvers growers have attained in their Carrot. Under their cultivation they raise from twenty to forty tons to the acre. This Carrot is of a rich, dark orange in color, very smooth and handsome, and from its length, is easier to dig than the Long Orange. It is a first-class Carrot for any soil.

Long Orange, or Long Surry. This is a standard variety, and in its various strains is doubtless more generally grown than any other kind. The chief objection to it is the depth to which it penetrates the ground, and hence the extra

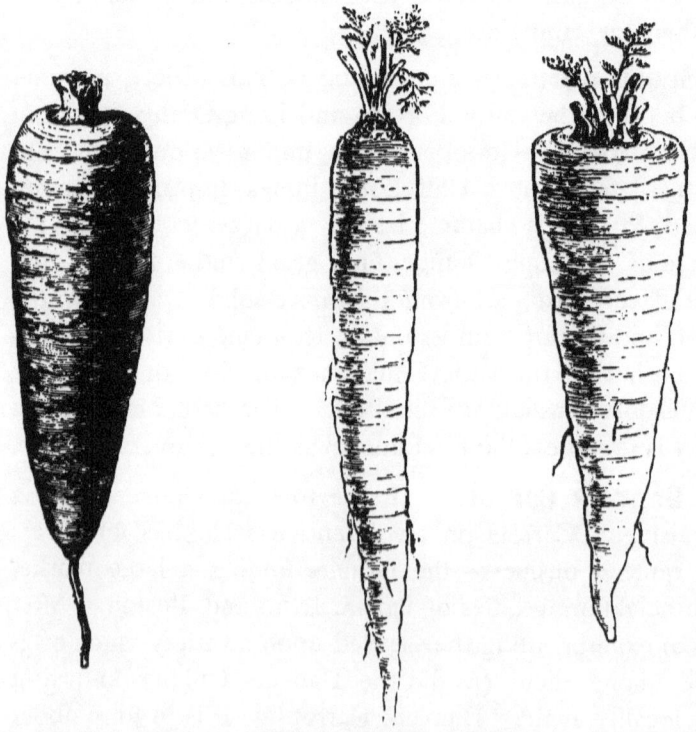

DANVERS CARROT   ALTRINGHAM.   IMPROVED LONG ORANGE.

work of digging it; while the end of the root which causes the extra work is of inferior quality when compared with the body, differing in this respect from the shorter varieties, which are of the same quality throughout. The heart is larger in proportion than in the shorter varieties, which is considered an objection. The keeping properties are excellent, and in this respect it is superior to the earlier kinds. On light soil the roots grow long, straight and make a fine show in the market.

Altringham. This is a Carrot of excellent quality for the table, the flesh being of a rich orange color, crisp and sweet, but as a cropper it is inferior to the Intermediate or Long Orange varieties, and hence is but little cultivated.

Large White Belgian. This is the largest of all varieties and will yield at least a quarter more than any other sort. The roots grow several inches out of ground, and all can be readily pulled by the hand. Analysis shows that it is nearly as sweet as the Mangold Wurtzel, rather sweeter than the Swede Turnip, and about two thirds as sweet as the Sugar Beet. The two objections to it

LARGE WHITE BELGIAN. are its color and its keeping properties; it being rather a poor keeper, while the color has made it a carrot for horses rather than cows. If farmers have but a small quantity of manure, the White Belgian is a good variety for them to raise for feeding early in the winter.

THE CULTIVATION, AND THE IMPLEMENTS NEEDED.

Just as soon as the young plants can be detected breaking ground, the prudent farmer will push the slide hoe, and have his boy weeders follow immediately after it on hands and knees. Boys that have had a little experience, with their nimble fingers can do more work than men, while their wages are only about half as much. On the sea-coast we hire boys who make a business of weeding, for from seventy-five cents to a dollar a day. The one great danger in hiring boys, is that careless ones are apt to break off the weeds instead of pulling them up by the roots. To ascertain their comparative faithfulness, it is well to quietly mark a few rows of the different weeders, at their first weeding, and by the time for the second weeding the difference between a good and a bad boy will be very plainly visible.

Don't accept that theory of the shiftless man, that it is well to have the weeds grow pretty tall before the first weeding, that the plants may be protected from the sun. I have noticed that oftentimes those who act on this theory give over their weeding, and plough up the bed before they have half finished it. Promptness in the first hoeing and weeding is exceedingly important in the management of all root crops, and it is where the great mistake is apt to be made in their cultivation.

There are a few implements that are specially needed in the cultivation of root crops, and of these every wise farmer will get the

SLIDE HOE.

very best attainable. These implements are the Seed Sower, the Hand Weeder, the Slide Hoe, the common Wheel Hoe, and one for weeding both sides of a row at the same time. Of these there are a great many varieties, each of which are more or less popular among a class of growers. The engravings illustrate such as are in use in my own section of country, where root culture forms a very important part of the agriculture of farmers. Both the slide and the wheel hoe, for rapid work, far surpass the common hand hoe, while they cut up the weeds equally clear. The wheel hoe is used until the tops of the crops become so large as to be in the way,

WHEEL HOE.

when the slide hoe takes its place. Each should be two inches narrower than the space between the rows. A slide hoe is an amazing handy implement about a farm for many uses other than between the rows of root crops. A new class of implements have been introduced within a few years which, to a degree, supersede the use of the common wheel or slide hoe, though there is yet a valuable sphere for each of them; I refer to the weeders which cut each side of the row at the same time

GOODWIN'S WHEEL HOE.

I have tested every variety of these and have thus far found none do such good, practical work as the homeliest looking one of them all, viz.: the Goodwin wheel hoe. These hoes which take each side of the row at once cannot safely be made to go over the ground as fast as those designed for use between the rows, but working close home to the growing crop, they save a large portion of the cost of hard weeding. Of seed drills there are a dozen or more in the market, several of which I have used on my farms. I prefer Matthew's over all others. Among other advantages it can be relied upon to

MATTHEW'S SEED SOWER.

drop almost any variety of small seed, while it is a good coverer, and having a roller attached, it packs the earth over the seed, which, as every farmer knows, tends to keep the moisture in and thus hastens their germination.

NOYES' WEEDER

The hand weeder is an excellent little implement to facilitate the laborious work of weeding, especially when the surface is baked and therefore rather hard on the fingers.

GATHERING AND STORING THE CROP.

One of the greatest outlays attending the raising of Carrots is in the gathering and topping of the crop. The common process of digging with a fork and throwing into piles to be afterwards topped is laborious and costly. The labor and consequent cost may be greatly lessened by first cutting off the tops by a sharp shovel, spade or common hoe, or a slide hoe which has been weighted by a piece of lead pipe, or some similar heavy article, slid down the handle and fastened where that unites with the hoe. Should a slice be taken off the tops of the roots it will do no harm, as Carrots differ in this respect from other roots, in that, when the tops are cut they are not apt to rot; indeed, some practice cutting off a slice of the root when topping, to keep them from sprouting so readily when stored.

Let the crop remain out as late as it can be risked without freezing; and if they are in good growing condition this will be well towards November, in the latitude of central New England, and even into the first week of that month in the milder temperature of the sea-coast. Roots not fully matured will keep better than those fully ripe when dug, on the principle that the varieties of apples we call "winter" apples are simply those kinds that do not ripen on the tree,—they are not winter apples, because they are Baldwins, or Greenings, for these same kinds in the South where the ripening season is longer, are Fall apples. If the carrots have been planted too early they will ripen before digging and be apt to prove poor keepers, besides losing the advantage of October weather which is the carrot month, doing more for the weight of the late planted crop than all the season besides.

Rake the tops off the bed but do not waste them for they are highly relished by animals, and if the carrots are harvested when they ought to be, to keep well, that is, when in good growing condition, there will be a great weight of tops, sometimes as high as a quarter of the weight of roots; and this mass of green fodder, coming at a time when the fields are usually bare of grasses, will prove very valuable and acceptable food for the cows. The common way of gathering the crop, by loosening with spades or forks and then pulling out by the tops, throwing into heaps or scattering over the ground and afterwards topping with a knife, is a long and costly job. An improvement on digging is to run a plough close to the row and then pull out as many as possible by hand and dig up the remainder. Still a better course particularly when the Danvers variety is grown, is, beginning in the middle of the piece, to run a subsoil plough close home to the roots, when, if run sufficiently deep it will lift the carrots a little out of the ground. Follow with forks or hoes, and draw the roots inward on the ploughed portion, so far as to give room for the horse to walk. Let the roots remain a few hours scattered over the surface, when in picking up and tossing them into carts or baskets, any earth adhering will be jarred off. In storing, one fact must be borne in mind; that carrots will heat, sprout and rot, under circumstances in which Mangolds would keep sound and uninjured. I have several times lost quantities when buried in the ground where Mangolds and common table Beets, under precisely the same conditions, have kept perfectly sound. If the crop is to be fed at once, they may be dumped into the cellar or barn floor in the most expeditious way without reference to the depth of the heap; but if to be fed into the winter, then all depth of the heap above two and a half to three feet means a proportionate increase of danger of heating, sprouting and rotting, and so much greater care to air the cellar in cool, dry weather. I need hardly state that cellars for keeping carrots and all roots should be free from standing water, and as cool as possible without actually freezing. If the bottom is damp, then put down a rough flooring. When the roots are large they will keep sufficiently better to pay for the extra trouble, if they are piled "heads and points" to the height of two and a half feet, with a slight space for air between the piles. If there are not cellar conveniences for storing the entire crop, with a good protection of hay under and around them, a few tons may be stored, for early feeding, in the barn, provided it is a warm one.

RAISING CARROTS WITH ONIONS.

I transfer from my Treatise on Onions, a paragraph relative to growing carrots with onions.

The plan of raising carrots with onions is considered a great improvement by many who have adopted it, as the yield of carrots is thought to be clear gain, diminishing but little or none the yield of onions. Carrots are planted in two ways; one by sowing them in drills between every other row of onions, and the other, which is considered an improvement, called the Long Island plan, by planting the onions in hills from seven to eight inches from center to center, dropping a number of seed in each hill, and from the first to the twelfth of June planting the carrot seed, usually by hand, between these hills in two rows, then skipping one, and thus on through the piece. The onions, as they are pulled are thrown into every third row, the carrots being left to mature. By this method from two to six hundred bushels of carrots are raised per acre in addition to the usual crop of onions. More manure is required for the two crops than for the onions alone.

The machine used for sowing in drills has two boxes attached to the axle at equi-distance from the wheels; there are three or four holes in the axle that communicate with the seed in the boxes, and as these holes pass under the boxes they are filled with seed, and as they turn the seed are dropped into the earth. Screws are sunk into the holes, which can be sunk more or less at pleasure, and the quantity of seed which the holes will contain is thus graded.

The machine should first be tested and so regulated that on a barn floor it will drop from eleven to twelve seed from each hole. When so regulated, on using in the field it will drop but from seven to twelve, owing to the more uneven motion.

MARKETING AND FEEDING.

In the cities there is a large market for carrots as feed for horses, it being very generally accepted that a few given daily or every other day, aids the digestion of grain-fed animals, adds to the gloss of the hair, and are of special medicinal value. The largest, smoothest and darkest orange colored roots sell the best in the market. The price varies all the way from ten to twenty dollars a ton of 2000 pounds, depending in part on the value of hay. Where the quantity fed daily is small a large knife or a shovel will answer to cut them up in pieces of suitable size; but if the quantity amounts to several bushels daily, then a root-cutter will be needed. There are two classes of these, one for sheep, and the other for large stock, the essential difference being that those designed to cut roots for sheep cut into smaller pieces. Of those designed to cut roots for large stock, the Whittemore machine is as good a machine as any, having a capacity to cut up a bushel in about half a minute. Among farmers there is much unnecessary fear about the danger of animals choking while feeding on apples, potatoes and roots. For the last ten years I have fed to my cows not far from three hundred tons of squashes, potatoes and roots, (mostly squashes) and never yet lost an animal or had any very serious trouble from choking. My habit is to feed them while quietly in their stalls, with a division board between the feed of each. All cases of choking that have come to my notice have occurred *where the animal was suddenly disturbed while eating.* There is a great difference of opinion as to how many roots can be fed to stock daily without injuring them. The proportion will depend somewhat on the constitutional peculiarities of individual cows, but when the bowels are all right the appetite of the animal is probably the safest guide. I have had a large and extended experience in feeding squashes to milch cows,—the Boston Marrow, Hubbard and other varieties; beginning with half a bushel to each animal, I increase the quantity until the daily consumption has averaged a hundred pounds a day to each. Under such heavy feeding, after a while their appetites clog somewhat, but I am inclined to the opinion that, beginning with a moderate feed, they would soon readily eat seventy-five pounds daily with a relish, for as long a period as they might last. When feeding Carrots or any roots, the most economical method is to give meadow or salt hay, with a small quantity of flax-seed or cotton-seed meal. The effect of the roots and these rich meals is to give to these inferior varieties of hay, the nutritious value of the best upland English.

MANGOLD WURTZELS.

What is a Mangold Wurtzel? A number of years ago I raised a piece of Early Turnip Beet seed in a very isolated location; there was not another piece of Beet seed growing within half a mile, at the least. A good deal of the seed wasted, as is usual when the seed is allowed to ripen well on the stock before cutting. From this waste seed thousands of young plants sprang up, many of which survived the winter, by the help of the protection of chickweed and snow. They had got so far along when ploughing time came, I left the piece unploughed, thinning them out that they might produce early beets. As the season advanced a good many of them pushed seed shoots and ripened a crop of seed. Some of the seed I gathered and the next season planted it to see what it would produce. The crop was "everything;" all the way from a nice, dark colored Early Turnip Beet, through different sizes, colors and forms, up to a light-fleshed Mangold Wurtzel. As the original Beets were a very pure Turnip Beet, and during several years of careful cultivation for seed purposes had shown no admixture with any other variety, the experiment proved either that the coarse variety of Stock Beet, which we call Mangold Wurtzel are but sports from our fine-grained table Beets, or that the Beet class are sports from Mangolds,—most probably the former.

Mangold Wurtzels differ from table Beets in their general coarseness of structure, and the larger size to which they grow, the elements which enter into the composition of each being the same in kind. I have grown an ordinary Turnip Beet to weigh twenty-three pounds, and of the size of a half bushel measure. At times, on rich, friable soil, the Long Blood Beet will attain to large proportions, but when led by such results to attempt to get equal weight with Mangolds, under first-rate conditions, the experiment, with me, has uniformly failed. Still, when quality is wanted, in the fattening of hogs for instance, I am not certain but that the food obtained from an acre of the large variety of table Beets, may not be more than that obtained from an equal acre in Mangolds.

What is a Sugar Beet? The term "Sugar Beet" is an unfortunate one, as the word "Sugar" had already been appropriated to express the sweet flavor of the varieties of Beets raised for table use, while the word Beet is strictly a misnomer, the vegetable Sugar Beet being in reality a Mangold Wurtzel. A generation ago our fathers used the term "Sugar" as a familiar designation for any sweet variety of beet raised for table use, and at the present by the great majority of the public the term is still so used. As the new industry of manufacturing sugar from the beet grew on the continent of Europe, seedsmen were called upon to supply for commerce seed of the best variety for this purpose. It was necessary that this variety should be as free as possible from all coloring substance as this would, as a matter of course, give a stain to the juice, and impose on the manufacturer the labor of purifying it. The ones at first selected were the long, white Mangold Wurtzels, and these were called the "Sugar" Beet in commercial parlance. These white Mangolds were not entirely white, the portion that grew above ground being usually colored a light green by exposure to the sun's rays; it became therefore an object for the manufacturer to still improve on them to the end that all the coloring should be eliminated. The intelligence and enterprise of the seedsmen of Europe responded to this want, and in the course of a few years two prominent varieties were produced, that have nearly completely satisfied it,—one of these was sent out by the estimable house of Vilmorin Andrieux & Co., of Paris, and is named "Vilmorin's New Improved White," and the other "White Imperial Extra," by the distinguished German house of Ernest Benary.

These improved Sugar Beets of commerce grow nearly entirely under ground, and when grown these beets define themselves to be the Mangold variety, by the coarser structure of the root, the stouter ribs and the greater coarseness of the leaves, which spring in larger masses directly from the crown, than is the case with beets for the table.

The moral of all this for my farmer friends is, that if you want a beet for table use do not order "Sugar Beet" or you will be very likely to find a Mangold growing in your garden, a return, but not a recompense for the sweat and toil of the husbandman.

VARIETIES.

About twenty varieties are catalogued by seedsmen, many of which are but strains of the same kind, bearing the name of the grower, who by careful cultivation has endeavored to improve it. Classified by form they come under three classes, viz. :—the long, the round and the ovoid or intermediate varieties. Classified by color we have the red or scarlet, the pink, the yellow or orange, and the white varieties.

The Long Varieties.—Among the more prominent of these are the Ox Horn, the common Long Red, Sutton's

Imperial, Norbiton Giant, Long Egyptian, Carter's Improved, the Long Yellow, and the Silesian varieties of Sugar Beet. The Ox Horn is a very crooked growing variety, as its name would imply, with a small diameter in proportion to its great length. Growing almost wholly out of ground it curves about so in the row as to be decidedly in the way, is apt to break when pulled and in addition to these defects, storing very badly, it is not in any way desirable. The Norbiton Giant, Carter's Mammoth Long Red, Sutton's Imperial, and Long Elvethan are improvements over the common Long Red in a greater uniformity in their habit of growth, their size, and a less liability to grow hollow at the top at the advanced stage of growth.

The Round Varieties.—In these are included the common Red and Yellow Globe, with some of the under-ground varieties of the Sugar Beet.

Ovoid are either red or yellow in color and are intermediate in form between the long and the round kinds.

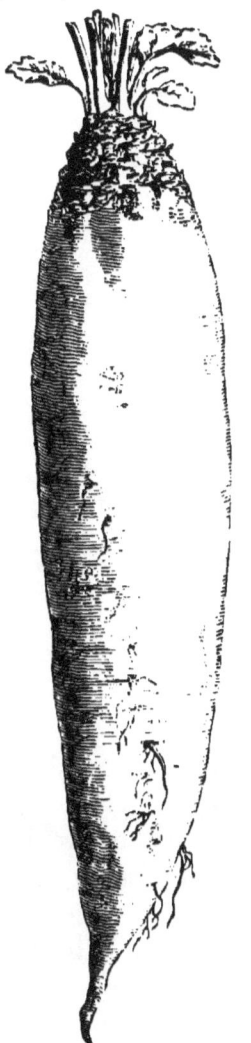

LONG RED MANGOLD.

What Kinds to Grow.— In this country the Long Red are the most popular, particularly the Norbiton Giant variety. While travelling in England, Ireland and France, for inquiry and observation, I found that the round and ovoid varieties were more generally cultivated than the long sorts. In my experience the ovoid varieties incline to grow smoother than the long kinds and hence are likely to bring up less earth with them, which on heavy soil is a matter of some

OVOID MANGOLD.

moment. I think of the two kinds the yellow, under the same circumstances, makes the larger root. The long varieties pile better in the cellar, while the round or ovoids cut up rather more readily, appear less inclined to rot at the top, and are firmer fleshed. The globe and ovoid varieties appear to be best adapted to hard and shallow soils, and of these the Yellow Globe and Ovoid are especially valuable, as they are better keepers than most sorts and remain sound, without sprouting, until late into the spring, and with special care may be kept even into the summer season.

The long Silesian varieties of Sugar Beet vary from each other only in the color of the part exposed above ground,—being green, grey or red. The kind introduced to the American public a few years ago, under the name of Lane's Improved American Sugar Beet, is a strain of the Long White Mangold. The improved varieties of Germany and France yield about double the per centage of sugar that is found in the common Mangold, in some crops the proportion being as high as thirteen per cent. This would make the Sugar Beets of double the value of Mangolds for stock, but unfortunately, the roots under like conditions of cultivation, average but half the weight of Mangolds.

As this treatise is about roots as food for stock, the cultivation of beet for the manufacture of sugar is not within its sphere, yet I must express surprise that with the experience of Germany and France to draw from and our own inventive skill and enterprise to add to it, we have not as yet made marked advance in this department of manufacturing industry. The average percentage of sugar found in analysis of beets grown in this country is exceptionably high. Land free from alkalies, of unbounded fertility, readily accessible, being attainable at almost nominal cost, it is a standing puzzle why we do not follow the example of other countries and raise our own sugar rather than import it. Perhaps the conundrum will be solved yet by some associate enterprise among our farmers, similar to that which gave

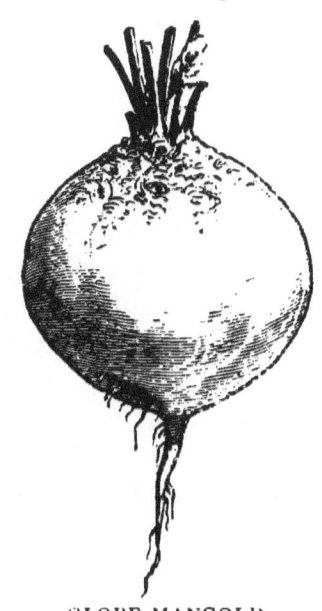

GLOBE MANGOLD.

birth to our cheese factory system; the inducement in this case being the home market that the sugar factory would afford for unlimited areas of beets, while the refuse pulp would enable them to increase greatly the number of their neat stock, to the advantage of the manure pile and enlargement of their area of tillage.

THE SOIL AND ITS PREPARATION.

In the matter of soil, Mangold Wurtzels will accept a greater latitude than any other root; thriving on every variety, all the way from light loam to muck, and from that to as strong a clay as is sufficiently friable for tillage. Muck (properly drained) and a strong loam are best suited to develop pounds of crop. Though the crop grown in the lighter soil is not so great, it is much sweeter than when grown on heavy soil, and when extraordinary quantities of manure have been applied, some of the heaviest crops on record have been grown on light loam. The great crop of Mr. Fearing of Hingham, of over sixty tons to the acre, was raised on a sandy loam. Some years ago I took a purchaser into the field where two lots of Mangolds were growing; he selected at once the large roots on the low land. I asked him to taste a slice of those on the upland, when he at once changed his preference. As a rule it will be found that those grown on warm, upland soil are decidedly the sweeter and this fact has an important bearing on the feeding value of the crop.

If the soil is in good heart for a foot in depth, plough it to that depth before putting on the manure. After putting on the manure, if coarse, it will be well to cut it up with Randall's wheel-harrow before ploughing under. After cross ploughing the manure four or five inches beneath the surface the aim should be to make a good seed bed by getting the surface level and the soil light and fine. On most soils this can be accomplished by a liberal use of the wheel-harrow followed by a fine-toothed smoothing harrow and that by a plank drag. An old barn door will sometimes answer for this, but as it is an excellent implement on the farm it will be well to have one. It should be about three feet wide and six long, with one side about ten inches high, meeting the bottom at an angle of forty-five degrees; the planks had better overlap slightly, as they will the better break the lumps of earth. The team is to be hitched to the turned up side, and the driver is to stand on the drag, driving it sideways over the land. The effect of such a drag in breaking up lumps and generally pulverizing the soil, will be found to be much superior to that of any roller. Should the soil be of such a character or in such a condition that the harrow and drag process will not make a good seed bed, there remains no resource other than to prepare it as for onions, by raking over the entire surface.

THE MANURE AND ITS APPLICATION.

The kind and quantities of food needed to grow any vegetable is found by an analysis of that vegetable. Having thus learned the kind and quantity needed for any crop, the next step of the wise farmer will be to ascertain what manures contain the necessary constituents and which of these contain them in the cheapest form. A little knowledge of Chemistry, in its application to manures, is of incalculable value to the husbandman and no amount of experience and traditionary knowledge can serve as a substitute for it. I believe that it is in this direction that the great advance in agriculture will be made, and were there no other argument for Agricultural colleges the fact that they are prepared to give thorough instruction in this one department would be a sufficient reason for their existence, and for their liberal patronage by their several states. Prof. Voelcker, an excellent authority in everything that pertains to chemistry, in its application to agriculture, gives the following table as the average composition of the ash of the principal root crops.

AVERAGE COMPOSITION OF THE ASH OF ROOTS.

	Number of Analyses.	Potash.	Soda.	Lime.	Magnesia.	Oxide of Iron.	Phosphoric Acid.	Sulphuric Acid.	Silica.	Chlorine.
Turnips.	38	49.8	7.3	11.7	2.6	0.9	10.3	11.8	1.2	5.1
Swedes.	7	38.9	14.0	19.8	4.2	0.3	10.4	13.7	1.9	4.2
Mangolds.	12	45.6	18.4	5.9	4.8	0.3	8.3	3.7	4.0	9.9
Sugar Be...	40	43.1	10.4	6.4	9.5	1.0	14.4	4.7	3.8	2.3
Carrots.	10	37.0	22.7	10.9	5.2	1.0	11.2	6.9	2.0	4.9
Parsnips.	4	45.7	2.7	15.7	6.0	1.3	15.8	5.6	2.4	4.0

LEAF ASH.

Turnips.	37	27.6	5.1	33.2	2.6	2.0	7.3	13.1	3.5	7.7
Swedes.	3	21.9	12.3	30.2	3.2	2.0	6.4	10.6	4.3	11.0
Mangolds.	4	25.5	23.3	10.4	9.7	1.2	5.4	7.2	3.3	17.8
Sugar Be...	7	21.9	16.6	19.5	18.1	1.3	7.3	7.9	3.1	5.7
Carrots.	7	17.6	18.2	32.1	3.9	3.0	3.8	3.2	5.2	8.9

This table shows us that the Mangolds require the mineral ingredients of manure in the following order, when arranged with reference to their importance:—Potash, Soda, Chlorine, Lime, Phosphoric Acid, Magnesia, Sulphuric Acid, Silica. In addition to these minerals other substances enter into the composition of Mangolds, the most important of which is Nitrogen. Barn-yard manure contains about all the elements needed by vegetation, but not always in the right proportion, therefore, when applying it, it is always profitable to know the proportions of the minerals which enter into crops that the deficiency may be supplied from other sources. It is perhaps hardly necessary to say that unleached wood ashes and the German Potash Salts, Sulphate and Muriate, are the cheapest sources for Potash at present known, while Soda and Chlorine are most cheaply obtained from the waste salt of the fisheries. Of this I shall have more to say presently when treating of salt as an auxiliary fertilizer. Lime is most cheaply obtained from the common Carbonate of Lime of the mason, either water or air slacked, and this usually contains more or less of Magnesia. The great source of Phosphoric Acid is the bones of animals or coprolites, by which is meant the fossilized bones and dung of extinct animals; Sulphuric Acid is most cheaply obtained from Plaster, which is Sulphate of Lime.

Some hold great benefit is derived by the crop of the following year, from ploughing under the leaves as soon as the roots are topped; the value of this is just what the analyses of our table shows. The large crops reported as raised in this country, have been raised on soil ranging from light to a friable clay loam and have received all the way from eight to fifteen cords of barn-yard manure to the acre. In some instances this has been all ploughed in; in others half spread broadcast and ploughed in and the other half put in the furrows. When coarse and unfermented I would advise a deep ploughing of it under, in the Fall As with Carrots, other waste substances can be used as substitutes for barn-yard manure, care being taken, either that such waste substances are specially rich in Potash, Soda and Chlorine, or that these substances be added. The equivalents given are roughly estimated under the article treating of the manure for Carrots and will be sufficient for practical purposes; I therefore make no further allusions to these cheap wastes as sources for manure, further than to mention that sea manures are specially rich in potash and soda.

Of all roots Mangolds are the rankest feeders, removing more plant food from the soil than any other root crop. The crop of Mr. Albert Fearing, of Hingham, Mass., was sixty tons of roots, and if the tops were in the usual proportion, of about one-third, they weighed twenty tons more, giving the enormous yield of eighty tons of green food from one acre of ground. The crop raised on Deer Island, in Boston harbor, was about seventy tons to the acre; with a like proportion of tops the total yield must have been over a hundred tons. In the sewage farms of England eighty tons of roots have been raised on an acre of ground. Fearing applied fifteen cords of manure to his acre of ground; of the quantity applied to the Deer Island crop I regret I have not the data at hand.

If the mere bulk alone was to be aimed at in the crop, the problem would be a very simple one, but there are three points to be considered: first, how to get a crop that shall be great in bulk and at the same time give us the second desirable point, viz.: *ripeness*, and thus insure the third desirable point, viz.: *the highest percentage of sugar it is possible for the roots to acquire.*

This matter of the value of Mangolds, for feeding purposes, being in about the same proportion as the sugar present, though appertaining to that part of this Treatise which treats of "Feeding to Stock," yet has so direct a bearing on the manuring of the crops that I will take it up at this place. The recent researches of that distinguished chemist, Prof. Voelcker of England, than whom there is no better authority, has thrown much light on the question of manure in its application to this crop. The Professor takes the position that the nutritious value of roots is in proportion to the amount of dry matter in them, and that the percentage of sugar present coincides with that of dry matter, the proportion of sugar rising or falling with the percentage of dry matter in the roots. That the feeding value does not depend on the proportion of nitrogen they contain, is proved theoretically, by the fact that the percentage is very much higher in the early stages of growth, before the crop is matured, than it is later in the season, while in the experiments of Mr. Lawes in feeding sheep, the lot containing the most nitrogen in the way of nutrition gave the poorest results.

Assuming with Prof. Voelcker that bulk should not be sought at a disproportionate sacrifice of sugar in the crop, and that certain soils and certain manures and certain methods of cultivation are more favorable than others to the development of this desirable proportion. I present extracts from his valuable article on "Root Crops as affected by Soil and Manures."

"Land highly manured with rich dung from the fattening boxes or stables, induces luxurious and vigorous growth in root crops, and, as is well known, has a tendency to develop over-luxuriance in the tops. This is the case more particularly if the dung is derived from fattening beasts, liberally supplied with oil-cake and artificial food, rich in nitrogenous constituents. If the Autumn turns out fairly dry and warm, the roots in highly manured land continue to grow vigorously, the bulbs swell to a large dimension, and if the weather in September and October continues warm and dry, a heavy weight, and fairly ripe roots, result from the liberal use of rich dung. But should the Autumn be cold and wet, too liberal an application of good, well-rotten dung is apt to maintain the luxuriant tops in a vigorous, active-growing condition, at a period of the year when the crop has to be taken up, and the result is an immature root crop, of a low feeding value. Although the bulbs may be of a good size, they turn out, when grown under such conditions, watery, deficient in sugar, and not nearly as nutritious as they would have been had a more moderate dressing of dung been put upon the land. The main cause of the immature condition and low-feeding quality of Mangolds grown with an excessive quantity of rich dung is the comparatively large amount of ammonial and nitrogenous constituents in the dung; for numerous field experiments have shown that the peculiar tendency of ammonia salts, and of readily available nitrogenous substances is to induce luxuriant leaf-development and vigorous and prolonged growth, which results frequently in a more or less immature condition of the roots. There is thus danger of over-manuring crops; and the desire to produce heavy crops of Mangolds not unfrequently leads practical men not to appreciate sufficiently this danger. It is quite true Mangolds are very greedy feeders, and no doubt some soils will swallow up almost any amount of dung; but at the same time it has to be borne in mind that all land is not alike, and that there are many naturally rich clay loams containing immense stores of plant food, which requires only to be brought into play by good cultivation in order to become available to plants. I am much inclined to think that it is a mistake to manure soils of the latter description too liberally with dung, even for Mangolds, and that in many cases a more economical result, and certainly a better quality of Mangolds, although not so heavy a crop, would be given, if, instead of all the enormous dressings of dung which are often applied to that crop, the land were manured in Autumn with only half the quantity of dung, and the seed drilled in with three to four cwt. of superphosphate or dissolved bones, which manures, as we shall see presently have a tendency to produce early maturity in roots. We frequently hear of complaints that Mangolds scour, or do not keep well. Complaints of this kind are only the expressions in other words for the immature condition of the roots, and in many cases the cause of this undesirable condition has to be sought in the excessive amount of ammonial or nitrogenous constituents which are applied to the Mangolds in the shape of heavy dressings of dung. The same remarks apply with equal force to the exclusive and too abundant use of Peruvian guano, sulphate of ammonia and nitrogenous manures in general. The special effect of all ammonial and nitrogenous manures in general, as already stated, is to produce luxuriant leaf development, to induce prolonged and vigorous growth, resulting in an immature and watery condition of the bulbs.

Large roots, generally speaking, are far less nutritious than better matured roots of a moderate size. For illustration of this fact I quote the following comparative analyses:

	Water.	Nitrogenous Constituents	Sugar, Pectine, &c.	Crude Fibre.	Ash.
MANGOLDS 9 lbs.	91.85	1.34	2.86	2.54	1.41
" 7 1-2 lbs.	89.48	1.24	3.95	4.51	.82
" 4 lbs.	89.77	0.73	7.68	.89	.93
" 1 to 2 lbs.	86.90	0.61	10.51	1.07	.91

Small Mangolds approach Sugar Beets in composition, whilst large Sugar Beets are hardly better than common Mangolds, and monster beets are even less nutritious than well-matured Mangolds of fair average size. Monster roots, as is well known, are always very watery, poor in sugar, and almost useless for feeding purposes.

Big Berkshire beets,—one weighing 16 pounds and the other 12 1-4 pounds,—contained only 3.89 or 4 per cent. of sugar respectively, and in round numbers as much as 91 1-2 per cent. of water. This high percentage of water is accompanied by a larger amount of albuminous compounds and of mineral matter, than the proportions in roots, containing very much more solid feeding matter. A large amount of albuminous matter and of ash, indeed indicates immaturity and poverty in sugar, a characteristic of big, excessively manured roots.

"Generally speaking, all nitrogenous manure, either should not be used at all, or only sparingly, for roots, on stiffish land, and all soils which contain a good deal of clay, are naturally cold and unfavorable to a vigorous and rapid growth. On the other hand, raw, or better still, dissolved Peruvian guano is an excellent manure for root crops upon light land, which, like most productive sandy soils and friable turnip loams, favors the quick and vigorous growth of roots, and is conducive to early maturity.

"Nitrate of soda has the same general effect upon root crops as nitrogenous manures, but it appears to be more energetic in its action, and, on the whole, to be a useful addition to home manures, and to increase the produce in roots more considerably than salts of ammonia. Its effect is specially marked upon mangolds, and, to my knowledge, heavy crops of mangolds have been produced upon rather light land by 1 1-2 per cent. of Nitrate of Soda, two cwt. of common salt, sown broadcast, and four cwt. of dissolved bones drilled in with the seed.

"Potash salts in some field experiments which I have tried in different parts of the country, have shown that Potash has a decidedly beneficial effect upon root crops, on poor, sandy soils; while on the majority of land, and notably upon clays or clay loams, or soils in a good agricultural condition, Salts of Potash do not increase the produce. The special effect of superphosphates, dissolved bones and similar phosphatic manures, is to produce early maturity; and hence phosphatic manures are employed in practice very largely, and with much benefit, by root growers. In free-growing, light soils, it is desirable either to use dissolved bones in addition to half dressing of farm-yard manure, as a manure for roots, or to spread broadcast 2 or 3 cwt. of salt, or 2 cwt. of guano and 1 cwt. of nitrate of soda and 2 cwt. of common salt, and to drill with the seed 3 to 4 cwt. of dissolved bones. On the heavier description of soils it is preferable to use mineral superphosphate for roots, especially if the land has been dressed in Autumn with a moderate quantity of dung."

SALT AS AN AUXILIARY MANURE.

It will be seen by the table of analysis of roots, that the Mangold has in it a remarkably large percentage of Chlorine and Soda, the roots yielding respectively 9.9 and 18.4, while the tops give, 17.8 and 23.3. Salt being a combination of Chlorine and Soda, known to chemists as Chloride of Sodium, must therefore be a valuable auxiliary manure for Mangolds, that is, one to be used in connection with other manures. Practice proves what chemistry indicates. Prof. Voelcker tells us that "salt tends to check over-luxuriance in the tops, while it prolongs the period of active growth. In consequence of this specific action it may be employed with benefit as an auxiliary manure upon light land, in quantities not greater than five bushels to the acre." Mr. Lewis, of New York, believes that by scattering over the surface, when the Mangolds develop the fourth leaf, four or five bushels of the refuse of the Syracuse salt works, which is about equal parts of salt and plaster, he has increased his crop ten tons to the acre. Mr. Lewis finds that salt tends to prevent a disease which sometimes attacks the leaves, known as "rust. He states that it can be obtained at the works for about $3.50 per ton. Prof. Voelcker believes it would be injurious rather than beneficial on heavy land.

The quantity to be applied to the acre as given by practical growers, varies from four to twenty-five bushels. The effect is not always the same; one season the increase may be very striking and the next, under the same application, not be perceptible, the cause of which is not very clear, though it appears to give better results in dry seasons than in wet. The most striking effect from the application of large quantities, in my experience, has been on the borders of meadow land. A number of years ago I manured in the furrow with refuse herring bait, salt and all, just as taken from the fish barrels. The crop of Mangolds grown from this manuring was one of the largest and smoothest I ever raised. The next season the land was planted to Oats. In the Fall, while laying a heap of this oat straw in the barn, I chanced to use one as a tooth-pick. It tasted as though it had been pickled; thinking it was the result of some accident, I took another; that also was salt. This aroused my curiosity and on examination I found farther, to my great surprise, that all the straw tasted as though it had been dipped in pretty strong brine. Certainly this tremendous salting, over and above what the crop of Mangolds could use, to all appearance, had not lessened the bulk of roots. On meadow land, Mr. Ware of this town, thinks that in a dry season he doubled his crop by the application of refuse salt, at the rate of twenty-five bushels to the acre. In purchasing waste salt for this or any other agricultural crop, it is best to get the dirtiest lot possible, for this dirt is the waste of the fish on which it has been used, and consists mostly of fish scales, which for manuring purposes is decidedly the most valuable part of the fish. For this reason the waste from salted herring is probably the most valuable of all. Under the open platforms where fish are dried, in sea-port towns, and from which there is always some dripping, the rankest of grass grows. Salt lessens the proportion of sugar in the roots.

PLANTING THE SEED AND TENDING THE CROP.

Our ground being now ready the next step is to plant it. How much seed shall we need and how far apart shall we have the rows? From four to six pounds of seed is the usual quantity, the higher figures evidently allowing for a considerable waste, while with hand planting even the smaller amount may be decreased. As to the proper distance between the rows, practical growers will give various replies;—18, 20, 22, 24, 30 inches. The thirty inch men are those who expect to depend on the cultivator to do about all their weeding, and are willing to prepare and spare more ground, with the object of having less weeding. That the crop does not require so much room to yield the greatest bulk, is shown by the experience of other cultivators, who have raised from forty to over sixty tons to the acre, with their rows from eighteen to twenty-two inches apart, while the greatest crop on record, viz. :—of over eighty tons to the acre, was raised with the rows twenty-four inches apart.

Planting on ridges is often advised, but as far as I have observed, those who begin this way generally change to the system of level culture as they advance in experience. The only advantages I have found in the system of ridge cultivation have been that the Mangolds appear to grow with fewer roots, and are rather more easily weeded. These advantages in practice are more than off-set by the extra labor of making the ridges and preparing them for planting. Mangold seed is apt to come up badly. In France, where land is cut up into small areas and labor is cheap, one would expect to find as little waste as possible, but while traveling there I noted in their fields that the Mangolds were quite scattering. Mangold seed, like those of beets, are enclosed in a porous shell which itself is usually called the seed. By cracking these "seeds" the real seed will be found within, at the angles, from one to four in number, and when broken, if fresh, appear as white as flour. One reason why a portion of the seed fails to vegetate, is, I infer, from the quantity of moisture necessary to reach and swell the encased seed. For this reason, if planted during dry spells, care should be taken to get them down to a good depth, say an inch and a half deep, and then to pack the fine earth closely over them so that it may hold the moisture. Any machine, therefore, that is used for planting should have a good roller. To facilitate and hasten the vegetation, some cultivators practice soaking the seed, by pouring on water when almost at a scalding temperature, and letting the seed remain in it from thirty-six to forty-eight hours, being careful to keep it where the water will not fall below blood heat, then rolling plaster or dry soil, until it is sufficiently dry to drop readily from the machine.

Some prefer to plant by hand, believing that the greater certainty of getting the seed up and the greater regularity of the plants in the row is more than an off-set to the additional labor. In doing this some growers will drop the seed on the surface by the machine, and then follow and push them under to the depth requisite, with the thumb and finger; others use a strip of plank about four inches wide and three feet in length, on the under side of which are inserted wooden pins, every seven inches, the pins being one and a quarter inches in diameter and projecting two inches. The holes having been made, the seed are dropped in, and covered by the hand. In my own experience I rely on Mathew's seed drill, and find but few blank places after the plants are up, provided the weather is not too dry. Where blanks are found they may be profitably filled by transplanting the young Mangolds, care being taken to break off the tops of the larger leaves, and also to loosen the ground a little when planting them. If a time just after a shower is selected, the result will be very satisfactory. The transplanted roots when gathered in the Fall will usually be found with several small roots in place of a single tap root.

All root crops require prompt and thorough attention in the matter of weeding, and to lessen this costly department of labor they should not be raised on land abounding in the seed of weeds. Mangolds will require two or three hand weedings, besides as many slidings with the scuffle or wheel-hoe. If too thick they should be thinned rather early in their growth, for I have oftentimes noticed that if this is left until the roots begin to develop, those left standing are apt to be dwarfed. It is best to give two thinnings. The plants should be left from ten to twelve inches apart; the crop of eighty tons was thinned to twelve inches apart, and as the roots are more apt to grow coarse and prongy, and with less sugar in them, when far apart, I am inclined to ten or twelve inches as far enough. The object aimed at should be, as Prof. Voelcker has shown, to get the weight in many roots of medium size rather than in fewer roots of large size.

GATHERING AND STORING THE CROP.

Unlike other roots, the keeping qualities of Mangolds are destroyed by a temperature low enough to but little more than freeze the surface of the ground. In the late Fall when the growth is about completed, these much exposed roots have but few leaves to protect them and hence, where freezing weather is feared, the provident farmer will always give them the benefit of the doubt. If he is so unfortunate as to have his crop injured, let him at once get the most he can out of them, in the way of food, for though the injury at first may appear to be but trivial, the part frozen will become first corky and afterwards turn black, and ultimately rot. If but slightly frozen the frost may be taken out by at once covering the roots temporarily with earth, but such roots must be fed early or they will rot. Where the globe or ovoid varieties are grown, on land where they pull hard they may be lifted by running a subsoil plough with care. In pulling these, or any roots that are to be topped on the field, don't do, as is usually done, either scatter them on the surface, without any system, or throw them into heaps, as in either way the cost of removing the tops is increased. If thrown in piles the tops become more or less intermingled, and the small amount of extra labor thereby caused in topping each individual root becomes great in the aggregate, when thousands are handled. Still it oftentimes happens that the weather takes a sudden, unexpected turn, threatening too low a temperature for the safety of the crop; under such circumstances the question is how to get it out of danger in the most expeditious way possible. The quickest way is to pull and throw into heaps, *roots in, tops out*, by which arrangement, should there be considerable of a freeze up, the tops would shield the roots. To protect them still more effectually earth may be shovelled over the heaps, so as barely to cover them, and when protected in this way they may be allowed to remain quite awhile awaiting the leisure of the farmer. Here let me say that this plan of protection will not answer for all crops, as I have learnt with Cabbages, to my sorrow, for when covered up this way, but for a few days, when taken out they will be found to be almost cooked by the great heat which they have developed.

In gathering all roots the great object is to have as few handlings as possible, hence, if the tops are not twisted off as the Mangolds are pulled, they should be laid in rows, tops in and roots out, four or more rows being put in one. It will be best to have two hands work together, and so make two of these rows, leaving a small passage-way between them, the roots being on the inside. Now let the topper follow with a large and sharp knife, and lop off the leaves to his right and left as he goes being careful to so top the roots that each individual leaf will fall separately, which means that he is not to cut the top of the root itself, for unlike Carrots, Mangolds so cut are apt to decay when stored. For economical work the knife should be a large and somewhat heavy one, the blade eight or nine inches in length. A small grit stone for the use of each of the hands engaged in topping any kind of roots is always a good investment; it saves running to the barn for an occasional touch on the grindstone.

If the roots are to be marketed they will need to be left awhile to have the earth on them dry, that it may fall off when loading, but if for use on the farm it will be rather of an advantage, as it will help keep them from wilting. The portion of the crop to be fed before Spring should be stored as near to the place of feeding as possible. The great object should be to keep them sufficiently covered and cool to prevent wilting. As all the beet family are good keepers, there need be but a small per cent. of loss. Store them in a cool, rather moist cellar, provided it has no standing water. The heap may be three or four feet in depth, and should be covered with earth that is rather moist than otherwise, to prevent evaporation. The long varieties may be piled cordwood fashion. Those to be fed after Spring opens can be kept in a pit, dug in gravelly soil, on a hill-side, or where there is no danger from standing water; the pit may be three or four feet in depth, and be filled to the surface. In covering there are two methods: one, to throw the earth directly on the roots, and the other to first cover them with cornstalks, or some dry, coarse litter before throwing on the earth. In practice I find that when the litter is used the roots in immediate contact with it are apt to mould, more or less, and be affected with a dry rot, though it is an excellent plan to throw over coarse litter up to severe freezing weather. Which ever course is pursued it is best not to throw on more at first than is sufficient to barely cover them, and to add the remainder, making a covering of about two feet in depth in all; to which is to be added a foot of coarse hay as the weather becomes cold. The process of thatching with straw and so piling that there shall be a roof-like slant to the heap, with furnace-like ventilators opening from it at intervals, I have never found necessary in actual practice, the elevation of the earth above the bed being a sufficient water shed, while the cold nature of the root prevents heating. Rats are the great enemies of root pits. I have had galleries cut by these vermin through a bed of roots, utterly destroying them for seed purposes. The best way of killing them in my experience, has been to drop a little arsenic on buttered bread and put it conveniently near their holes, but so far hidden that no neighbors dog would be likely to suffer by it.

FEEDING THE CROP.

Besides arguments which are of weight for cultivation of all kind of roots, there are special ones for the raising of Mangolds. The vast bulk of yield exceeds that of any annual crop, as high as eighty tons of roots having been raised to the acre on the sewerage farms of England, and when to this is added the weight of leaves that such a crop would carry, it will be safe to say that a hundred tons have been given to the acre. Taken as a whole the Mangold has less enemies and is less apt to fail than any other root. Compared with the Turnip family, it has several marked advantages, being more reliable in dry seasons and less liable to disease; and in flesh-forming, heat-giving and fat-producing elements it surpasses it. While the Turnip family cannot be raised repeatedly on the same land, indeed on most soil can be raised only at intervals of three or four years, Mangolds can be raised many years in succession, as Mr. Mechi, the distinguished English agriculturist, has proved by raising sixty tons per annum on the same tract of land of six acres area, for six successive years. They will keep longer in good condition than any other root, under favorable circumstances even as late as July. Experiments in feeding steers made with care, proved that while a ton of Mangolds increased their weight sixty-five pounds, a ton of Swede increased their weight but forty-eight pounds, equal quantities of hay having been fed in each experiment. Other experiments have established about the same proportionate value between these two roots, though the general result was not as favorable. Mangolds, like fruit, undergo a ripening change after they are gathered, and until this is effected they are not in the best condition for feeding. The ripening process for the most part consists in a change of starch into sugar, and makes the Mangolds both more healthful and more nutritious food. Before this change is effected they are apt to scour stock if fed to any degree liberally. The time when this chemical change takes place will depend on the degree of ripeness of the crop when stored; and this, as has been clearly shown is affected by both the soil on which they grew and the manure with which they were fed; other conditions equal, those grown on upland ripen earlier than those on lowland, while rank manures tend to prolong the period of growth and crops so grown, come into condition for feeding later in the season. In England, a common practice is to begin feeding the Mangolds at Christmas, while in this country the middle of January is considered early enough. Experiments carefully made have proved that when fed to fattening animals they should follow and not precede Turnips. It is a good rule in feeding this as with other roots or tubers, to begin with a small quantity and gradually increase the amount up to the limit which the appetite of the cow, her general health and the tale of the milk pail indicates. Every farmer who feeds a dairy needs a root cutter. There are several of these in the market, some designed for sheep only, which cut the roots into small pieces, others for neat cattle, while some manufactured by our Canada neighbors can be arranged to cut for either class of stock. As good a one as I know of for stock purposes, cheapness, durability and effectiveness combined, is one sometimes known as the Whittemore machine, of which I present an engraving. This machine is capable of cutting about two bushels a minute. Experiments in England have shown that 59 pounds of cooked Mangolds are equal to 70 of uncooked; but that meat made from steamed food wastes more when boiled. Leaves of Mangolds should be fed with care as they are more apt to scour than those of any other root. The reason of this is that they contain comparatively a large quantity of a poisonous acid, known by chemists as "oxalic" acid, the same that is developed in Rhubarb leaves, when slightly wilted, and which sometimes causes death when such leaves are eaten as "greens."

WHITTEMORE CUTTER.

The practice sometimes followed in Europe, of feeding the leaves of the growing crop, where labor is very cheap, is thought to pay, as the leaves are gathered just as they begin to drop from their upright position and when their usefulness as nourishers of the root have ended. But with labor as cheap as may be, there is no economy in this, for, aside from the deleterious effects to animals, when fed too liberally, by actual experiment it has been found that the wear and tear to the crop, incidental to the plucking of these leaves by an average farm hand, injures it more than the value of the leaves after they are gathered.

Were it not for the enormous bulk that an acre will produce in roots when compared with its yield in hay or grain, there would be a serious argument against the growing of them to any extent beyond what might be needed for medicinal purposes, in the fact that the manure made from them is of so low a value; and the practical weight of this argument would grow in proportion as farmers acquire a knowledge of the most important department of farming. To most farmers a cord or load of manure of cow or horse is a cord or load of equal value; now this is far, very far from being the fact, as will be seen by the following table which I take from the *Scientific Farmer*, compiled by the celebrated Mr. Lewis, who, by his careful experiments has laid the agricultural world under lasting obligation. In this table a ton of English hay is taken as the standard, and were all the manure saved, both solid and liquid, from a ton of each of these varieties of food, the ingredients at the market value of the Ammonia, Potash and Phosphoric Acid would be worth as follows:—

Hay,	$10.00
Clover Hay,	15.00
Oat Straw,	4.50
Wheat Straw,	4.16
Barley Straw,	3.50
Decorticated Cotton Seed Cake,	43.33
Linseed Cake,	30.66
Malt Dust,	28.33
Malt,	10.50

Oats, - - - - - - 11.50
Wheat, . - - - - 11.00
Indian Corn, - - - - - 10.50
Barley, - - - - - - 9.83
Potatoes, - - - - 2.33
Mangolds, - - - - - - 1.66
Swedes, - - - - - - 1.41
Turnips, (common,) - - - - 1.33
Carrots, - - - - - - 1.33

 This table is very suggestive in many ways:—by it we see that there are varieties of food, the manure from which is worth more than the cost of the food itself. In its application to the feeding of Mangolds, it at a glance suggests the wisdom of feeding at the same time a portion of something richer and more concentrated. By so doing the quality of the manure is vastly improved and the crops will not be slow to discover it. There is still another reason for feeding these rich foods while using roots; it enables the armer to feed with profit his straw or inferior varieties of hay. Says Prof. Stockhardt, "the full benefit to animals derivable from feeding roots is secured only when the proper proportion of substances rich in nitrogen are fed with them; accordingly, about two pounds of oil-cake should be fed with each hundred pounds of beet root, or other foods may be substituted in the same proportion as they are rich in nitrogen."

 Recent researches have determined a fact of great value to agriculture; that to get the most profitable results from food the Albuminoid and Carbohydrate elements should bear a certain proportion to each other, and that while a decrease in either of them from this proper proportion means insufficient food, and a consequent loss of flesh, fat or milk, an excess of either means money wasted. The proportion for cows that are dry and oxen when not at work, is about, one of Albuminoids to eight of Carbohydrates; for oxen at work and cows in milk, one of Albuminoids to from four to six of Carbohydrates.

The following table taken from Prof. Johnson's excellent work, "How Crops Grow," gives the proportion of the Albuminoids, Carbohydrates and other elements in roots and tubers.

ROOTS AND TUBERS.	Water.	Organic Matter.	Ash.	Albuminoids	Carbohydrates.	Crude Fibre.	Fat, &c.
POTATO.	95.0	24.1	0.9	2.0	21.0	1.1	0.3
JERUSALEM ARTICHOKE.	80.0	18.9	1.1	2.0	15.6	1.3	0.5
KOHL-RABI.	88.0	10.8	1 2	2.3	7.3	1.2	0.2
FIELD BEETS, (3 lbs. weight).	88.0	11.1	0.9	1.1	9.1	0.9	0.1
SUGAR BEETS. (1 to 2 lbs.)	81.5	17.7	0.8	1.0	15.4	1.3	0.1
RUTA BAGAS, (about 3 lbs.)	87.0	12.0	1.0	1.6	9.3	1.1	0.1
CARROT, (about 1-2 lb.)	85.0	14.0	1.0	1.5	10.8	1.7	0.2
GIANT CARROT. (1 to 2 lbs.)	87.0	12.2	0.8	1.2	9.8	1.2	0.2
TURNIPS.	92.0	7.2	0.8	1.1	5.1	1.0	0.1
PARSNIP.	83.3	11.0	0.7	1.6	8.4	1.0	0.2
PUMPKIN.	94.5	4.5	1.0	1.3	2.8	1.0	0.1

To give the tables necessary to develop this interesting subject to its full capacity, would be altogether beyond the scope of my little treatise. I will refer my readers to the appendix of that excellent work by Prof. Johnson, "How Crops Grow."

THE COST OF THE CROP.

An average crop of Mangolds may be set down at 22 tons. To grow this crop would cost the farmer who depends on barn manure mainly, about as follows:—

DEBTOR.

Ploughing twice, harrowing and dragging,	$9.00
Seed,—4 lbs.,	3.00
Planting,	1.00
Sliding, weeding and thinning crop,	16.00
Gathering, topping and storing,	12.00
Manure, and handling of 7 cords,	56.00
Refuse salt, 16 bushels, at $1.25 per hogshead,	2.50
Interest, taxes and wear and tear of implements and teams,	15.00
Total cost,	$114.50

CREDITOR.

By crop of 22 tons roots, at $8.50 per ton,	$187.00
" tops,—4 tons, at $5.00,	20.00
" value of manure left in soil,	14.00
	$221.00
	114.50
Balance,	$106.50

In the above estimate I have assumed most of the labor to be by boys, who at hand weeding, if they are reliable, can get over the ground faster than men. I have made no allowance for the cost of cutting up the roots when feeding, as this does not belong under this head. Should the land be old the item of weeding would have to be increased one-half. The salt I have priced at its cost along the sea-coast. I have estimated the value of the crop at the average value of several years past, while the manure charge is much higher than it should be where farmers have access to the fertilizing wastes of great cities.

Now, if instead of being contented with a crop of 22 tons to the acre, the farmer strives for double that quantity, he will get it by additional expense in but two directions, viz.: his manure bill and the cost of gathering and storing. If we now double the cost of each of the latter, and credit the results with double the crop, which every practical farmer who has had experience in root culture will allow is but reasonable, we shall have the following results :—

Extra cost of crop of 44 tons over one of 22 :

Manure,—7 cords,	$56.00
Gathering, topping and storing,	12.00
	$68.00

Now adding the credit side we shall have for

Extra 22 tons roots,	$187.00
6 tons tops,	30.00
Value of manure left in ground,	14.00
	$231.00
Deduct extra cost,	68.00
Profits cleared,	$163.00

In other words, by investing $68.00 for six months, we clear $163.00, which, as any farmer boy can figure, is at the rate of about five hundred per cent. a year. Mr. Fearing of Hingham, with the same amount of manure raised over sixty tons to the acre, and the instances are numerous where over forty tons have been the crop when even a less quantity has been used. Can any farmer who has accumulated a small surplus of money do better than invest it in manure? There is altogether too much money, for the prosperity of their farming, invested by farmers in Savings Banks. These banks pay from six to seven per cent. on money, but here is an instance where an investment made in manure pays over four hundred per cent. Merchants don't do so foolish a thing as to put their earnings into Savings Banks. No; they invest in their business and so keep it and its money making capacity under their own control; when will farmers be as wise and become their own bankers? Let me remark that the farmer who is so wise as to attempt to get the most from his land will do well to follow Prof. Voelcker's advice and drill in four or five hundred weight of some good phosphate, to the acre, in place of the same value in stable manure.

In the above estimates of the value of Mangolds we have assumed that the farmer sold his crop. Now it is true of this as of every other crop that the farmer can use on his premises, that it is of more value to him than the general market price indicates.

Under this head an intelligent farmer of large experience writes:—

"From experiments made in feeding beets, their practical value has been made to range from 13 to 20 cents per bushel, with hay at twenty dollars per ton. An exact estimate of the practical value of beets for cattle food, is a difficult matter, as it is now, and ever will be, hid from mortal ken. The improved condition of the cow, (when fed to cows during the winter,) her increased usefulness during the entire season, her lessened liability to sickness and disease which high feeding with any one of the different kinds of grain induces, her lengthened lease of life, her evident satisfaction and perfect contentment, which is so plainly manifested while eating her daily ration of roots, are each and every one legitimate items to be taken into the account in estimating the practical, the actual value of beets as food for dairy stock.

"After carefully looking at the subject in all its bearings, so far as my experience has given me opportunity to do so, I have come to the conclusion that beets for cattle food are well worth fully as many cents per bushel as good hay is worth dollars per ton, without taking into consideration the increase of the manure; and that the average cost, when stored in the cellar or put into pits, with every item of expense included, need not exceed eight cents per bushel."

I will close my little treatise by remarking that while I cannot expect to have exhausted so prolific a subject, yet I hope and trust that it may prove of value as a guide and a stimulus to some of my many friends in the great community of farmers.

CABBAGES:

HOW TO GROW THEM.

A PRACTICAL

Treatise on Cabbage Culture,

GIVING FULL DETAILS ON EVERY POINT, INCLUDING KEEPING AND MARKETING THE CROP.

BY

JAMES J. H. GREGORY,

INTRODUCER OF THE MARBLEHEAD CABBAGES.

MARBLEHEAD:
MESSENGER STEAM PRINTING HOUSE.
1883.

Entered according to Act of Congress, in the year 1870, by
JAMES J. H. GREGORY,
At the Clerk's Office of the District Court of Massachusetts.

CONTENTS.

	Page.
Object of Treatise	3
The Origin of Cabbage	3
What a Cabbage is	4
Selecting the Soil	6
Preparing the Soil	7
The Manure	8
How to Apply the Manure	10
Making the Hills and Planting the Seed	12
Care of the Young Plants	16
Protecting the Plants from their Enemies	18
The Green Worm	21
Club or Stump Foot	22
Care of the Growing Crop	27
Marketing the Crop	28
Keeping Cabbages through the Winter	30
Having Cabbage Make Heads in Winter	36
Varieties of Cabbage	38
Early York	40
Large York	40
Early Oxheart	41
Early Sugar Loaf	41
Early Winnigstadt	41
Red Dutch	42
Red Drumhead	42
The Little Pixie	43
Early Schweinfurt, or Schweinfurt Quintal	44
Early Wakefield	45
Early Wyman	45
Premium Flat Dutch	46
Early Low Dutch	47
Stone Mason	47
Large Late Drumhead	48
Marblehead Mammoth	49
American Green Glazed	50

	Page.
Fottler's Early Drumhead	50
Bergen Drumhead	51
Cannon Ball	51
Savoy Cabbage	52
Drumhead Savoy	52
Pancalier	53
Early Ulm Savoy	53
Early Dwarf Savoy	54
Improved American Savoy	54
Golden Savoy	54
Norwegian Savoy	54
Victoria, Russian, Cape Savoys	55
Feather Stemmed Savoy	55
Large Brunswick Short Stem'd	55
Early Empress	55
Robinson's Champion Ox Drumhead	55
English Winnigstadt	55
Blenheim	55
Shillings Queen	55
Carter's Superfine Early Dwarf	55
Enfield Market Improved	56
Kemp's Incomparable	56
Fielderkraut	56
Ramsay's Winter Drumhead	56
Pomeranian Cabbage	56
Alsacian Cabbage	56
Marbled Burgogne	56
Early Dutch Drumhead	56
Cabbage Greens	56
Cabbage for Stock	58
Raising Cabbage Seed	61
Cooking Cabbage, Sour Krout, etc.	62
Cabbage Under Glass	64
Cold Frame and Hot Bed	66
Cauliflower, Broccoli, Brussels Sprouts, Kale and Sea Kale	68

OBJECT OF THIS TREATISE.

As a general yet very thorough response to inquiries from many of my customers about cabbage raising, I have aimed in this treatise to tell them all about the subject. The different inquiries made from time to time have given me a pretty clear idea of the many heads under which information is wanted; and it has been my aim to give this with the same thoroughness of detail as in my little work on Squashes. I have endeavored to talk in a very practical way, drawing from a large observation and experience, and receiving, in describing varieties, some valuable information from McIntosh's work, "The Book of the Garden."

THE ORIGIN OF CABBAGE.

Botanists tell us that all of the Cabbage family, which includes not only every variety of cabbage, Red, White, and Savoy, but all the cauliflower, broccoli, kale, and brussels sprouts, had their origin in the wild cabbage of Europe, (*Brassica oleracea,*) a plant with green, wavy leaves, much resembling charlock, found growing wild at Dover in England, and other parts of Europe. This plant, says McIntosh, is mostly confined to the sea shore, and grows only on chalky or calcareous soils.

Thus through the wisdom of the Great Father of us all, who occasionally in his great garden allows vegetables to sport into a higher form of life, and grants to some of these sports sufficient strength of individuality to enable them to perpetuate themselves, and at times to blend their individuality with that of other sports, we have the heading cabbage in its numerous varieties, the creamy cauliflower, the feathery kale, the curled savoy. On my own grounds from a strain of seed that had been grown isolated for years, there recently came a plant that in its structure closely resembled Brussels Sprouts, growing about two feet in height, with a small head under each leaf. The cultivated cabbage was first introduced into England by the Romans, and from there nearly all the kinds cultivated in this country were originally brought. Those which we consider as peculiarly American varieties, have only been made so by years of careful improvement on the original imported sorts. The characteristics of these varieties will be given farther on.

WHAT A CABBAGE IS.

If we cut vertically through the middle of the head, we shall find it made up of successive layers of leaves, which grow smaller and smaller, almost *ad infinitum*. Now if we take a fruit bud from an apple tree and make a similar section of it, we shall find the same structure. If we observe the development of the two, as Spring advances, we shall find another similarity (the looser the head the closer will be the resemblance),—the outer leaves of each will unwrap and unfold, and a blossom stem will push out from each. Hence we see that a cabbage is a bud, a seed bud, as all fruit buds may be termed, the production of seed being the primary object in nature, the fruit enclosing it playing but a secondary part, the office of the leaves being to cover, protect, and afterwards nourish the young seed shoot. The outer leaves which surround the head appear to have the same office as the leaves which surround the growing fruit bud, and that office closes with the first year, as does that of the leaves surrounding fruit buds, when each die and drop off. In my locality the public must have perceived more or less clearly the analogy between the heads of cabbage and the buds of trees, for when they speak of small heads they frequently call them " buds." That the close wrapped leaves which make the cabbage head and surround the seed germ, situated just in the middle of the head at the termination of the stump, are necessary for its protection and nutrition when young, is proved, I think, by the fact that those cabbages the heads of which are much decayed, when set out for seed, no matter how sound the seed germ may be at the end of the stump, never make so large or healthy a seed shoot as those do the heads of which are sound ; as a rule, after pushing a feeble growth, they die.

For this reason I believe that the office of the head is similar to and as necessary as that of the leaves which unwrap from around the blossom buds of our fruit trees. It is true that the parallel cannot be fully maintained, as the leaves which make up the cabbage head do not to an equal degree unfold, (particularly is this true of hard heads) ; yet they exhibit a vitality of their own, which is seen in the deeper green color the outer leaves soon attain, and the change from tenderness to toughness in their structure ; I think, therefore, that the degree of failure in the parallel may be measured by the difference between a higher and a lower form of organic life.

Some advocate the economy of cutting off a large portion of the heads when cabbages are set out for seed to use as food for stock. There is certainly a great temptation, standing amid acres of large, solid heads in the early Spring months, when green food of all kinds is scarce, to cut and use such an immense amount of rich food, which, to the inexperienced eye, appears to be utterly wasted if left to decay, dry, and fall to the ground; but, for the reason given above, I have never done so. It is possible that large heads may bear trimming to a degree without injury to the seed crop; yet I should consider this an experiment, and one to be tried with a good deal of caution.

SELECTING THE SOIL.

In some of the best cabbage growing sections of the country, until within a comparatively few years it was the very general belief that cabbage would not do well on upland. Accordingly the cabbage patch would be found on the lowest tillage land of the farm. No doubt, the lowest soil being the richer from a gradual accumulation of the wash from the upland, when manure was but sparingly used, cabbage would thrive better there than elsewhere,—and not, as was generally held, because that vegetable needed more moisture than any other crop. Cabbage can be raised with success on any good corn land, provided such land is well manured; and there is no more loss in seasons of drouth on such land than there is in seasons of excessive moisture on the lower tillage land of the farm. I wish I could preach a very loud sermon to all my farmer friends on the great value of liberal manuring to carry crops successfully through the effects of a severe drouth. Crops on soil precisely alike, with but a wall to separate them, will in a very dry season present a striking difference,— the one being in fine vigor, and the other "suffering from drouth," as the owner will tell you, but in reality from want of food.

The smaller varieties of cabbage will thrive well on either light or strong soil, but the largest drumheads do best on strong soil. For the *Brassica* family, including cabbages, cauliflowers, turnips, etc., there is no soil so suitable as freshly turned sod, provided the surface is well fined by the harrow, and it is well to have as stout a crop of clover or grass, growing on this sod when turned under as possible ; and I incline to the belief that it would be a judicious investment to start a thick growth of these by the application of guano to the surface sufficiently long before turning the sod to allow for its effects on the growth of the clover or grass. If the soil be very sandy in character, I would advise that the variety planted be the Winnigstadt, which in my experience is unexcelled for making a hard head under almost any conditions, however unpropitious. Should the soil be naturally very wet it should be underdrained, or stump foot will be very likely to appear, which is death to all success.

PREPARING THE SOIL.

Should the soil be a heavy clay, a deep Fall ploughing is best, that the frosts of Winter may disintegrate it; and should the plan be to raise an early crop, this end will be promoted by Fall ploughing, on any soil, as the land will thereby be made dryer in early Spring. In New England the soil for cabbages should be ploughed as deep as the subsoil, and the larger drumheads should be planted only on the deepest soil. If the season should prove a favorable one, a good crop of cabbage may be grown on sod broken up immediately after a crop of hay has been taken from it, provided plenty of fine manure is harrowed in. One great risk here is from the dry weather that usually prevails at that season, preventing the prompt germination of the seed, or rooting of the plants. It is prudent in such a case to have a good stock of plants, that such as die may be promp'ly replaced.

The manure may be spread on the surface of either sod or stubble land and ploughed under, or be spread on the surface after ploughing and thoroughly worked into the soil by the gang plough or cultivator. On ploughed sod I have found nothing so satisfactory as the class of wheel harrows, which not only cut the manure up fine and work it well under, but by the same operation can be made to cut and pulverize the turf until the sod is left not over an inch in thickness. To do the work thus thoroughly requires a yoke of oxen or a pair of stout horses. All large stones and large pieces of turf that are torn up and brought to the surface should be carted off before making the hills.

THE MANURE.

Any manure but hog manure for cabbage,—barn manure, rotten kelp, night soil, guano, phosphates, wood ashes, fish, salt, glue waste, hen manure, slaughter-house manure. I have used all of these, and found them all good when rightly applied. If pure hog manure is used it is apt to produce that corpulent enlargement of the roots known in different localities as "stump foot," "underground head," "finger and thumb." I have found barn manure on which hogs have run, two hogs to each animal, excellent. The cabbage is the rankest of feeders, and to perfect the larger sort a most liberal allowance of the richest composts is required. To grow the smaller varieties either barn-yard manure, guano, phosphates, or wood ashes, if the soil be in good condition, will answer; though the richer and more abundant the manure the larger are the cabbages, and the earlier the crop will mature.

To perfect the large varieties of drumhead—by which I mean to make them grow to the greatest size possible—I want a strong compost of barn-yard manure, with night soil and muck, and, if possible, rotten kelp. A compost into which night soil enters as a component is best made by first covering a plot of ground of easy access, with soil or muck that has been exposed to a winter's frost, to the depth of about eighteen inches, and raising around this a rim about three feet in height, and thickness. Into this the night soil is poured from carts built for the purpose, until the receptacle is about two-thirds full. Barn manure is now added, being dropped around and covering the outer rim, and if the supply is sufficient, on the top of the heap also, on which it can be carted after cold weather sets in. Early in Spring the entire mass should be pitched over, thoroughly broken up with the bar and pick where frozen, and the frozen masses thrown on the surface. In pitching over the mass work the rim in towards the middle of the heap. After the frozen lumps have thawed give the heap another pitching over, aiming to mix all the materials thoroughly together, and make the entire mass as fine as possible. A covering of sand thrown over the heap before the last pitching will help fine it.

To produce a good crop of cabbages with a compost of this quality, from five to twelve cords will be required to the acre. If the land is in good heart by previous high cultivation, or the soil is naturally very strong, five cords will give a fair crop of the small varieties; while, with the same conditions, from nine to twelve cords to the acre will be required to perfect the largest variety grown, the Marblehead Mammoth Drumhead.

Of the other kinds of manure named above I will treat farther under the head of

HOW TO APPLY THE MANURE.

The manure is sometimes applied wholly in the hill, at other times partly broadcast and partly in the hill. If the farmer desires to make the utmost use of his manure for that season, it will be best to put most of it into the hill, particularly if his supply runs rather short; but if he desires to leave his land in good condition for next year's crop, he had better use part of it broadcast. My own practice is to use all my rich compost broadcast, and depend on guano, phosphates, or hen manure in the hill. Let all guano, if at all lumpy, like the Peruvian, be sifted, and let all the hard lumps be reduced by pounding, until the largest pieces shall not be larger than half a pea, before it is brought upon the ground. My land being ready, the compost worked under and the rows marked out, I select three trusty hands who can be relied upon to follow faithfully my directions in applying so dangerous manure as guano is in careless or ignorant hands; one takes a bucket of it, and, if for large cabbage, drops as much as he can readily close in his hand, where each hill is to be; if for small sorts, then about half that quantity, spreading it over a circle about a foot in diameter; the second man follows with a pronged hoe, or better yet, a six-tined fork, with which he works the guano well into the soil, first turning it three or four inches under the surface, and then stirring the soil *very thoroughly* with the hoe or fork. Unless the guano (and this is also true of most phosphates), is faithfully mixed up with the soil, the seed will not vegetate. Give the second man about an hour the start, and then let the third man follow with the seed. Of phosphates I use about half as much again as of guano to each hill, and of hen manure a heaping handful, after it has been finely broken up, and, if moist, slightly mixed with dry earth. When salt is used, it should not be depended on exclusively, but be used in connection with other manures at the rate of from ten to fifteen bushels to the acre, applied broadcast over the ground, or thoroughly mixed with the manure before that is applied; if dissolved in the manure, better yet. Fish and glue waste are exceedingly powerful manures, very rich in ammonia, and if used the first season they should be in compost. It is best to handle fish waste, such as heads, entrails, backbones, and liver waste, precisely like night soil. "Porgy cheese," or "chum," the refuse after pressing out the oil from menhaden, and now sold extensively for manure, is best prepared for use by composting it with muck or loam, layer with layer, at the rate of a barrel to every foot and a half, cord measure, of soil. As soon as it shows some heat turn it, and repeat the process two or three times, until it is well decomposed, when apply. Glue waste is a very coarse, lumpy manure, and requires a great deal of severe manipulation if it is to be applied the first season. A better way is to compost it with soil, layer with layer, having each layer about a foot in thickness, and so allow it to remain over until the next season before using. This will decompose most of the straw, and break down the hard, tough lumps. In applying this to the crop, most of it had better be used broadcast, as it is apt at best to be rather too coarse and concentrated to be used liberally directly in the hill. Slaughter-house manure should be treated much like glue manure.

MAKING THE HILLS AND PLANTING THE SEED.

The idea is quite prevalent that cabbages will not head up well except the plants are started in beds and then transplanted into the hills where they are to mature. This is an error, so far as it applies to the northern states; —the largest and most experienced cultivators of cabbage in New England usually dropping the seed directly where the plants are to stand, unless they are first started under glass, or the piece of land to be planted cannot be prepared in season to enable the farmer to put his seed directly in the hill and yet give the cabbage time sufficient to mature. Where the climate is unpropitious, or the quantity of manure applied is insufficient, it is possible that transplanting may promote heading. The advantages of planting directly in the hill are a saving of time, avoiding the risks incidental to transplanting, and having all the piece start alike ; for when transplanted many die and have to be replaced, while some hesitate much longer than others before starting, thus making a want of uniformity in the maturing of the crop. There is also this advantage, there being several plants in each hill, the cut-worm has to depredate pretty severely before he really injures the piece ; again, should the seed not vegetate in any of the hills, every farmer will appreciate the advantage of having healthy plants growing so near at hand that they can be transferred to the vacant spaces with their roots so undisturbed that their growth is hardly checked. In addition to the labor of transplanting saved by this plan, the great check that plants always receive when so treated is prevented, and also the extra risks that occur should a season of drouth follow.

Some of our best farmers drill their seed in with a sowing machine, such as is used for onions, carrots, and other vegetable crops. This is a very expeditious way, and has the advantage of leaving the plants in rows instead of bunches, as in the hill system, and thus enables the hoe to do most of the work of thinning. It has also this advantage, each plant being by itself can be left much longer before thinning, and yet not grow long in the stump, thus making it available for transplanting, or for sale in the market, for a longer period.

The usual way of preparing the hills is to strike out furrows with a small, one-horse plough, as far apart as the rows are to be. As it is very important that the rows should be as straight as practicable, it is a good plan to run back once in each furrow, particularly on sod land where the plough will be apt to catch in the turf and jump out of line. A manure team follows, containing the dressing for the hills, which has previously been pitched over and beaten up until all the ingredients are fine and well mixed. This team is so driven, if possible, as to avoid running in the furrows. Two or three hands follow with forks or shovels, pitching the manure into the furrows at the distance apart that has been determined on for the hills. The manure is leveled with hoes, a little soil is drawn over it, and a slight stamp with the back of the hoe is given to level this soil and at the same time to mark the hill. The planter follows with seed in a tin box, or any small vessel having a broad bottom, and taking a small pinch between the thumb and fore finger he gives a slight scratch with the remaining fingers of the same hand, and dropping in about half a dozen seed covers them half an inch deep with a sweep of the hand, and packs the earth by a gentle pat with the open palm to keep the moisture in the ground and thus promote the vegetation of the seed. With care a quarter of a pound of seed will plant an acre, when dropped directly in the hills; but half a pound is the common allowance, as there is usually some waste from spilling, while most laborers plant with a free hand.

The soil over the hills being very light and porous, careless hands are apt to drop the seed too deep. Care should be taken not to drop the seed all in one spot, but to scatter them over a surface of two or three inches square, that each plant may have room to develop without crowding its neighbors.

If the seed is to be drilled in, it will be necessary to scatter the manure all along the furrows, then cover with a plough, roughly leveling with a rake.

Should the compost applied to the hills be very concentrated, it will be apt to produce stump foot; it will therefore be safest in such cases to hollow out the middle with the corner of the hoe, or draw the hoe through and fill in with earth, that the roots of the young plants may not come in direct contact with the compost as soon as they begin to push.

When guano or phosphates are used in the hills it will be well to mark out the rows with a plough, and then, where each hill is to be, fill in the soil level to the surface with a hoe, before applying them. I have in a previous paragraph given full instructions how to apply these. Hen manure, if moist, should be broken up very fine, and be mixed with some dry earth to prevent it from again lumping together, and then applied in sufficient quantity to make an equivalent of a heaping handful of pure hen manure to each hill. Any liquid manure is excellent for the cabbage crop; but it should be well diluted, or it will be likely to produce stump foot.

Cabbage seed of almost all varieties are nearly round in form, but are not so spherical as turnip seed. I note, however, that seed of the Improved American Savoy is nearly oval. In color they are light brown when first gathered, but gradually turn dark brown if not gathered too early. An ounce contains nearly ten thousand seed, but should not be relied upon for many over two thousand good plants, and these are available for about as many hills only when raised in beds and transplanted; when dropped directly in the hills it will take not far from eight ounces of the larger sorts to plant an acre, and of the smaller cabbage rather more than this. Cabbage seed when well cured and kept in close bags will retain their vitality four or five years; old gardeners prefer seed of all the cabbage family two or three years old.

When the plan is to raise the young plants in beds to be transplanted, the ground selected for the beds should be of rich soil; this should be very thoroughly dug, and the surface worked and raked very fine, every stone and lump of earth being removed. Now sprinkle the seed evenly over the bed and gently rake in just under the surface, compacting the soil by pressure with a board. As soon as the young plants appear, sprinkle them with air-slaked lime. Transplant when three or four inches high, being very careful not to let the plant get tall and weak.

For late cabbage, in the latitude of Boston, to have cabbages ready for market about the first of November, the Marblehead Mammoth should be planted the 20th of May, other late drumheads from June 1st to June 12th, provided the plants are not to be transplanted; otherwise a week earlier. In those localities where the growing season is later, the seed should be planted proportionally later.

CARE OF THE YOUNG PLANTS.

In four or five days, if the weather is propitious, the young plants will begin to break ground, presenting at the surface two leaves, which together make nearly a square, like the first leaves of turnips or radishes. As soon as the third leaf is developed, go over the piece, and boldly thin out the plants. Wherever they are very thick, pull a mass of them with the fingers and thumb, being careful to fill up the hole made with fine earth. After the fourth leaf is developed, go over the piece again and thin still more; you need specially to guard against a slender, weak growth, which will happen when the plants are too crowded. In thinning, leave the short-stumped plants, and leave them as far apart in the hill as possible, that they may not shade each other, or so interfere in growing as to make long stumps. If there is any market for young plants, thousands can be sold from an acre when the seed are planted in the hill; but in doing this bear in mind that your principal object is to raise cabbages, and to succeed in this the young plants must on no account be allowed to stand so long together in the hills as to crowd each other, making a tall, weak, slender growth,—getting " long legged," as the farmers call it.

If the manure in any of the hills is too strong, the fact will be known by its effects on the plants, which will be checked in their growth, and be of a darker green color than the healthy plants. Gently pull away the earth from the roots of such with the fingers, and draw around fresh earth; or, what is as well or better, transplant a healthy plant just on the edge of the hill. When the plants are finger high they are of a good size to transplant into such hills as have missed, or to market. When transplanting, select a rainy day, if possible, and do not begin until sufficient rain has fallen to moisten the earth around the roots, which will make it more likely to adhere to them when taken up. Take up the young plants by running the finger or a trowel under them; put these into a flat basket or box, and in transplanting set them to the same depth they originally grew, pressing the earth a little about the roots.

If it is necessary to do the transplanting in a dry spell, as usually happens, select the latter part of the afternoon, if practicable, and, making holes with a dibble or any pointed stick an inch and a half in diameter, fill these holes, a score or more at a time, with water; and as soon as the water is about soaked away, beginning with the hole first filled, set out your plants. The evaporation of the moisture below the roots will keep them moist until they get a hold. Cabbage plants have great tenacity of life, and will rally and grow when they appear to be dead; the leaves may all die, and dry up like hay, but if the stump stands erect and the unfolded leaf at the top of the stump is alive, the plant will usually survive. Some advocate wilting the plants before transplanting; others challenge their vigor by making it a rule to do all transplanting under the heat of mid-day. I think there is not much of reason in either course.

PROTECTING THE PLANTS FROM THEIR ENEMIES.

As soon as they have broken through the soil, an enemy awaits them in the small black insect commonly known as the cabbage or turnip fly, beetle, or flea. This insect, though so small as to appear to the eye as a black dot, is very voracious and surprisingly active. He apparently feeds on the juice of the young plant, perforating it with small holes the size of a pin point. He is so active when disturbed that his motions cannot be followed by the eye, and his sense of danger is so keen that only by cautiously approaching the plant can he be seen at all. The delay of a single day in protecting the young plants from his ravages will sometimes be the destruction of nearly the entire piece. Wood ashes and air-slaked lime, sprinkled upon the plants while the leaves are moist from either rain or dew, afford almost complete protection. The lime or ashes should be applied as soon as the plant can be seen, for then, when they are in their tenderest condition, the fly is most destructive. I am not certain that the alkaline nature of these affords the protection, or whether a mere covering by common dust might not answer equally well. Should the covering be washed off by rain, apply it anew immediately after the rain has ceased, and so continue to keep the young plants covered until the third or fourth leaves appear, when they will have become too tough to serve as food for this insect enemy.

A new enemy much dreaded by all cabbage raisers will begin to make his appearance at about the time the flea disappears, known as the cut-worm. This worm is of a dusky brown color, with a dark colored head, and varies in size up to about two inches in length. He burrows in the ground just below the surface, is slow of motion, and does his mischievous work at night, gnawing off the young plants close at the surface of the ground. This enemy is hard to battle with. If the patch be small, these worms can be scratched out of their hiding places by pulling the earth carefully away the following morning for a few inches around the stump of the plant destroyed, when the rascals will usually be found half coiled together. Dropping a little wood ashes around the plants close to the stumps is one of the best of remedies ; its alkaline properties burning his nose I presume. A tunnel of paper put around the stump but not touching it, and sunk just below the surface, is recommended as efficacious ; and from the habits of the worm I should think it would prove so. Late planted cabbage will suffer little or none from this pest, as he disappears about the middle of June. Some seasons they are remarkably numerous ; making it necessary to replant portions of the cabbage patch several times over. I have heard of as many as twenty being dug at different times the same season out of one cabbage hill. The farmer who tilled that patch earned his dollars. When the cabbage has a stump the size of a pipe stem it is beyond the destructive ravages of the cut worm, and should it escape stump foot has usually quite a period of growth free from the attacks of enemies. Should the season prove unpropitious and the plant be checked in its growth, it will be apt to become " lousy," as the farmers term it, referring to its condition when attacked by a small green insect known as aphidæ, which preys upon it in myriads ; when this is the case the leaves lose their bright green, turn of a bluish cast, the leaf stocks lose somewhat of their supporting powers, the leaves curl up into irregular shapes, and the lower layer turns black and drops off, while the ground under the plant appears covered with the casts or bodies of the insects as with a white powder. When in this condition the plants are in a very bad way.

Considering the circumstances under which this insect appears, usually in a very dry season, I hold that it is rather the product than the cause of disease, as with the bark louse on our apple trees ; as a remedy I advocate sprinkling the plants with air-slaked lime, watering if possible, and a frequent and thorough stirring of the soil with the cultivator and hoe. The better the opportunities the cabbage have to develop themselves through high manuring, sufficient moisture, good drainage, and thorough cultivation, the less liable they are to be "lousy." As the season advances there will sometimes be found patches eaten out of the leaves, leaving nothing but the skeleton of leaf veins ; an examination will show a band of caterpillars of a light green color at work, who feed in a compact mass, oftentimes a square, with as much regularity as though under the best of military discipline. The readiest way to dispose of them is to break off the leaf and crush them under foot. The common large red caterpillar occasionally preys on the plants, eating large holes in the leaves especially about the head. When the cabbage plot is bordered by grass land, in seasons when grasshoppers are plenty, they will frequently destroy the outer rows, puncturing the leaves with small holes and feeding on them until little besides their skeletons remain. In isolated locations rabbits and other vegetable feeders sometimes commit depredations. The snare and the shot gun are the remedy for these.

Other insects that prey upon the cabbage tribe, in

their caterpillar state, are the cabbage moth, white-line, brown-eyed moth, large white garden butterfly, white and green veined butterfly. All of these produce caterpillars which can be destroyed either by application of air-slaked lime, or by removing the leaves infested and crushing the intruders under foot. The cabbage-fly, father-long legs, the mellipedes, the blue cabbage fly, brassy cabbage flea, and two or three other insect enemies are mentioned by McIntosh as infesting the cabbage fields of England; also three species of fungi known as white rust, mildew, and *cylindrosporium concentricum;* these last are destroyed by the sprinkling of air-slaked lime on the leaves. In this country, along the sea coast of the northern section, in open ground cultivation there is comparatively but little injury done by these marauders, which are the cause of so much annoyance and loss to our English cousins.

THE GREEN WORM.

A new and troublesome enemy to the cabbage tribe has made its appearance within a few years, and spread rapidly over a large section of country, in a green worm. This pest infests the cabbage tribe at all stages of its growth; it is believed to have been recently introduced into this country from Europe, by the way of Canada, where it was brought in a lot of cabbage. It is probably the caterpillar of a white butterfly with black spots on its wings. In Europe, this butterfly is preyed on by two or more parasites, which keep it somewhat in check; but its remarkably rapid increase in this country, causing a wail of lamentation to rise in a single season from the cabbage growers over areas of tens of thousands of square miles, leads me to fear that it has reached this country without its attendant parasites.

Besides this green worm, there are found in Europe four varieties of caterpillar variously marked, all of which make great havoc among the cabbage tribe.

The remedies given as successful, by writers and correspondents, are sprinkling with clarified lime-water, dusting with air-slaked lime, hellebore, or Scotch snuff. An admixture of carbonate and carbolate of lime, dusted on, has proved a protective in this country. Chickens allowed to run freely among the growing plants, the hen being confined in a movable coop, if once attracted to them will fatten on them. This remedy might answer very well for small plots. Water heated to the temperature of 160° and applied at once, being sprinkled over the plants by a common brush broom, has proved a success; but at 180° it has injured the leaves. Large areas in cabbage, in proportion to their size are, as a rule, far less injured by insect enemies than small patches.

CLUB OR STUMP FOOT.

The great dread of every cabbage grower is a disease of the branching roots, producing a bunchy, gland-like enlargement, known in different localities under the name of club foot, stump foot, underground head, finger and thumb. The result is a check in the ascent of the sap, which causes a defective vitality. There are two theories as to the origin of club foot; one that it is a disease caused by poor soil, bad cultivation, and unsuitable manures; the other that the injury is done by an insect enemy, *Curculio contractus*. It is held by some that the maggots at the root are the progeny of the cabbage flea; this I doubt. This insect, "piercing the skin of the root, deposits its eggs in the holes, lives during a time on the sap of the plant, and then escapes and buries itself for a time in the soil."

If the wart or gland-like excrescence is seen while transplanting, throw all such plants away unless your supply is short; in such case carefully trim off all the diseased portions with a sharp knife. If the disease is in the growing crop, it will be made evident by the drooping of the leaves under the mid-day sun, leaves of diseased plants drooping more than those of healthy ones, while they will usually have a bluer cast. Should this disease show itself, set the cultivator going immediately, and follow with the hoe, drawing up fresh earth around the plants, which will encourage them to form new fibrous roots; should they do this freely the plants will be saved, as the attacks of the insect are usually confined to the coarse branching roots. Should the disease prevail as late as when the plants have reached half their growth, the chances are decidedly against raising a paying crop.

When the land planted is too wet, or the manure in the hill is too strong, this dreaded disease is liable to be found on any soil; but it is most likely to manifest itself on soils that have been previously cropped with cabbage, turnip, or some other member of the Brassica family.

Farmers find that as a rule *it is not safe to follow cabbage, ruta baga, or any of the Brassica family, with cabbage, unless three or four years have intervened between the crops;* and I have known an instance in growing the Marblehead Mammoth, where, though five years had intervened, that portion of the piece occupied by the previous crop could be distinctly marked off by the presence of club foot.

Singular as it may appear, old gardens are an exception to this rule. While it is next to impossible to raise in old gardens a fair turnip free from club-foot, cabbages may be raised year after year on the same soil with impunity, or at least with but trifling injury from that disease. This seems to prove, contrary to English authority, that club-foot in the turnip tribe is the effect of a different cause from the same disease in the cabbage family.

There is another position taken by Stephens in his "Book of the Farm," which facts seem to disprove. He puts forth the theory that " all such diseases arise from poverty of the soil, either from want of manure when the soil is naturally poor, or rendered effete by over-cropping." There is a farm on a neck of land belonging to this town which has peculiar advantages for collecting sea kelp and sea moss, and these manures are there used most liberally, particularly for the cultivation of cabbage, from eight to twelve cords of rotten kelp, which is stronger than barn manure and more suitable food for cabbage, being used to the acre. A few years ago, on a change of tenants, the new incumbent heavily manured a piece for cabbage and planted it; but as the season advanced stump foot developed in every cabbage on one side of the piece, while all the remainder were healthy. Upon inquiry he learned that by mistake he had overlapped the cabbage plot of last season just so far as the stump foot extended. In this instance it could not have been that the cabbage suffered for want of food, for not only was the piece heavily manured that year and the year previous, but it had been liberally manured through a series of years, and to a large extent with the manure which of all others the cabbage tribe delight in, rotten kelp and sea mosses. I have known other instances where soil naturally quite strong and kept heavily manured for a series of years has shown stump foot when cabbage were planted with intervals of two and three years between. My theory is that the *mere presence of the cabbage* causes stump foot on succeeding crops grown on the same soil. This is proved by the fact that where a piece of land in grass, close adjoining a piece of growing cabbage, had been used for stripping them for market, when this was broken up the next season and planted to cabbage, stump foot appeared only on that portion where the waste leaves fell the year previous. I have another instance to the same point, told me by an observing farmer, that on a piece of sod land on which he run his cultivator the year previous when turning his horse every time he had cultivated a row, he had stump footed cabbage the next season just as far as that cultivator went, dragging, of course, a few leaves and a little earth from the cabbage piece with it. Still, though the mere presence of cabbage causes stump foot, it is a fact that under certain conditions cabbage can be grown on the same piece of land year after year successfully, with but very little trouble from stump foot. In this town (Marblehead) though, as I have stated, we cannot on our farms follow cabbage with cabbage, even with the highest of manuring and cultivation, yet in the gardens of the town, on the same kind of soil, (and our soil is green stone and syenite, not naturally containing lime,) there are instances where cabbage have been successfully followed by cabbage on the same spot for a quarter of a century and more. In the garden of an aged citizen of this town, cabbages have been raised *on the same spot of land* for over half a century.

The cause of stump foot cannot therefore be found in the poverty of the soil, either from want of manure or its having been rendered effete from over-cropping. It is evident that by long cultivation soils gradually have diffused through them something that proves inimical to the disease that produces stump foot. I will suggest as probable that the protection is afforded by the presence of some alkali that old gardens are constantly acquiring through house waste which is always finding its way there, particularly the slops from the sink, which abound in potash. This is rendered further probable from the fact given by Mr. Peter Henderson, that, on soils in this vicinity, naturally abounding in lime, cabbage can be raised year following year with almost immunity from stump foot. He ascribes this to the effects of lime in the soil derived from marine shells, and recommends that lime from bones be used to secure the same protection; but the lime that enters into the composition of marine shells is for the most part carbonate of lime, whereas the greater portion of that which enters into the composition of bones is phosphate of lime. Common air-slaked lime is almost pure carbonate of lime, and hence comes nearer to the composition of marine shells than lime from bones, and, being much cheaper, would appear to be preferable.

An able farmer told me that by using wood ashes liberally he could follow with cabbage the next season on the same piece. An experiment of my own in this direction did not prove successful, where ashes at the rate of two hundred bushels to the acre were used; and I have an impression that I have read of a like want of success after quite liberal applications of lime. Still, it remains evident, I think, that nature prevents stump foot by the diffusing of alkalies through the soil, and I mistrust that the reason why we sometimes fail with the same remedies is that we have them mixed, rather than intimately combined, with the particles of soil.

As I have stated under another head, an attack of club foot is almost sure to follow the use of pure hog manure, whether it be used broadcast or in the hill. About ten years ago I ventured to use hog manure nearly pure, spread broadcast and ploughed in. Stump foot soon showed itself. I cultivated and hoed the cabbage thoroughly; then, as they still appeared sickly, I had the entire piece thoroughly dug over with a six-tined fork, pushing it as deep or deeper into the soil than the plough had gone, to bring up the manure to the surface; but all was of no use; I lost the entire crop. Yet, on another occasion, stable manure on which hogs had been kept, at the rate of two hogs to each animal, gave me one of the finest lots of cabbage I ever raised.

CARE OF THE GROWING CROP.

As soon as the young plants are large enough to be seen with the naked eye, in with the cultivator and go and return once in each row, being careful not to have any lumps of earth cover the plants. Follow the cultivator immediately with the hoe, loosening the soil about the hills. The old rule with farmers is to cultivate and hoe cabbage three times during their growth, and it is a rule that works very well where the crop is in good growing condi'ion; but if the manure is deficient, the soil bakes, or the plants show signs of disease, then cultivate and hoe once or twice extra. "Hoe cabbage when wet," is another farmer's axiom. In a small garden patch the soil may be stirred among the plants as often as may be convenient, it can do no harm; cabbages relish tending; though it is not necessary to do this every day, as one enthusiastic cultivator evidently thought, who declared that by hoeing his cabbages every morning he had succeeded in raising capital heads.

If a season of drouth occurs when the cabbages have begun to head, the heads will harden prematurely; and then should a heavy rain fall, they will start to make a new growth, and the consequence will be many of them will split. Split or bursted cabbage are a source of great loss to the farmer, and this should be carefully guarded against by going frequently over the piece when the heads are setting, and starting every cabbage that appears to be about mature. A stout pronged potato hoe applied just under the leaves, and a pull given sufficient to start the roots on one side, will accomplish what is needed. If cabbage that have once been started seem still inclined to burst, start the roots on the other side. Instead of a hoe they may be pushed over with the foot, or with the hand. Frequently, heads that are thus started will grow to double the size they had attained when about to burst.

MARKETING THE CROP.

When preparing for market cabbages that have been kept over Winter, particularly if they are marketed late in the season, the edges of the leaves of some of the heads will be found to be more or less decayed; do not strip such leaves off, but with a sharp knife cut clean off the decayed edges. The earlier the variety the sooner it needs to be marketed, for as a rule cabbages push their shoots in the Spring in the order of their earliness. If they have not been sufficiently protected from the cold, the stumps will often rot off close to the head and sometimes the rot will include the part of the stump that enters the head. If the watery looking portion can be cut clean out, the head is salable; otherwise it will be apt to have an unpleasant flavor when cooked.

As a rule, cabbages for marketing should be trimmed into as compact a form as possible; the heads should be cut off close to the stump, leaving two or three spare leaves to protect them. They may be brought out of the piece in bushel baskets, and be piled on the wagon as high as a hay stack, being kept in place by a stout canvas sheet tied closely down. In the markets of Boston, in the fall of the year they are usually sold at a price agreed upon by the hundred head; this will vary not only with the size and quality of the cabbage, but with the season, the crop, and the quality in market on that particular day. Within a few years I have known the range of price for the Stone Mason or Fottler cabbage equal in size and quality, to be from $3 to $17 per hundred; for the Marblehead Mammoth from $3 to $25 per hundred. Cabbages brought to market in the Spring are usually sold by weight or by the barrel, at from $1 to $4 per hundred pounds.

The earliest cabbages carried to market sometimes bring extraordinary prices; and this has created a keen competition among market gardeners, each striving to produce the earliest, a difference of a week in marketing oftentimes making a difference of one-half in the profits of the crop. Capt. Wyman, who controlled the Early Wyman cabbage for several years, sold some seasons thirty thousand heads, if my memory serves me, at pretty much his own price. As a rule, it is the very early and the very late cabbages that sell most profitably. Should the market for very late cabbages prove a poor one, the farmer is not compelled to sell them, no matter at what sacrifice, as would be the case a month earlier; he can pit them, and so keep them over to the early Spring market which is almost always a profitable one. In marketing in Spring it should be the aim to make sale before the crops of Spring greens become plenty, as these replace the cabbage on many tables. By starting cabbage in hot beds a crop of celery or squashes may follow them the same season.

KEEPING CABBAGES THROUGH THE WINTER.

In the comparatively mild climate of England, where there are but few days in the Winter months that the ground remains frozen to any depth, the hardy cabbage grows all seasons of the year, and turnips left during Winter standing in the ground are fed to sheep by yarding them over the different portions of the field. With the same impunity, in the southern portion of our own country the cabbages are left unprotected during the Winter months; and, in the warmer portions of the South they are principally a Winter crop. As we advance farther north, we find that the degree of protection needed is afforded by running the plough along each side of the rows, turning the earth against them, and dropping a little litter on top of the heads. As we advance still farther northward, we find sufficient protection given by but little more than a rough roof of boards thrown over the heads, after removing the cabbages to a sheltered spot and setting them in the ground as near together as they will stand without being in contact, with the tops of the heads just level with the surface.

In the latitude of New England, cabbages are not secure from injury from frost with less than a foot of earth thrown over the heads. In mild Winters a covering of half that depth will be sufficient; but as we have no prophets to foretell our mild Winters, a foot of earth is safer than six inches. Where eel grass can be procured along the sea coast, or there is straw or coarse hay to spare, the better plan is to cover with about six inches of earth, and when this is frozen sufficiently hard to bear a man's weight, (which is usually about Thanksgiving time) to scatter over it the eel grass, straw or coarse hay, to the depth of another six inches. In keeping cabbages through the Winter, three general facts should be borne in mind, viz.: that repeated freezing and thawing will cause them to rot; that excessive moisture or warmth will also cause rot; while a dry air, such as is found in most cellars, will abstract moisture from the leaves, injure the flavor of the cabbage, and cause some of the heads to wilt and the harder heads to waste. In the Middle States we have mostly to fear the wet of Winter, and the plan for keeping for that section should therefore have particularly in view protection from moisture, while in the northern States we have to fear the cold of Winter, and consequently our plan must there have specially in view protection from cold.

When storing for Winter, select a dry day, if possible sufficiently long after rainy weather to have the leaves free of water,—otherwise they will spout it on to you, and make you the wettest and muddiest scarecrow ever seen off a farm,—then strip all the outer leaves from the head but the two last rows, which are needed to protect it. This may be readily done by drawing in these two rows towards the head with the left hand, while a blow is struck against the remaining leaves with the fist of the right hand. Next pull up the cabbage, which, if they are of the largest varieties, may be expeditiously done by a potato hoe. If they are not intended for seed purposes, stand the heads down and stumps up until the earth on the roots is somewhat dry, when it can be mostly removed by sharp blows against the stump given with a stout stick. In loading do not bruise the heads. Select the place for keeping them in a dry, level location, and if in the North a southern exposure, where no water can stand and there can be no wash. To make the pit, run the plough along from two to four furrows, and throw out the soil with the shovel to the requisite depth, which may be from six to ten inches; now if the design is to roof over the pit, the cabbages may be put in as thickly as they will stand; if the heads are solid they may be either head up or stump up, and two layers deep; but if the heads are soft, then heads up and one deep, and not crowded very close that they may have room to make heads during the Winter. Having excavated an area twelve by six feet, set a couple of posts in the ground midway at each end, projecting about five feet above the surface; connect the two by a joist secured firmly to the top of each, and against this, extending to the ground just outside the pit, lay slabs, boards or poles, and cover the roof that will be thus formed with six inches of straw or old hay, and if in the North throw six or eight inches of earth over this. Leave one end open for entrance and to air the pit, closing the other end with straw or hay. In the North close both ends, opening one of them occasionally in mild weather.

When cabbages are pitted on a large scale this system of roofing is too costly and too cumbersome. A few thousand may be kept in a cool root cellar, by putting one layer heads down, and standing another layer heads up between these. The common practice in the North, when many thousands are to be stored for Winter and Spring sales, is to select a southern exposure having the protection of a fence or wall if practicable, and turning furrows with the plough throw out the earth with shovels to the depth of about six inches; the cabbages, stripped as before described, are then stored closely together, and straw or coarse hay is thrown over them to the depth of a foot or eighteen inches. Protected thus they are accessible for market at any time during the Winter. If the design is to keep them over till Spring, the covering may be first six inches of earth, to be followed as cold increases, with six inches of straw, litter, or eel grass. This latter is my own practice, with the addition of leaving a ridge of earth between every three or four rows to act as a support and keep the cabbages from falling over. I am also careful to bring the cabbages to the pit as soon as pulled, with the earth among the roots as little disturbed as possible, and should the roots appear to be dry, to throw a little earth over them after the cabbages are set in the trench. The few loose leaves remaining will prevent the earth from sifting down between the heads, and the air chambers thus made answer a capital purpose in keeping out the cold, as air is one of the best non-conductors of heat. It is said that muck soil when well drained is an excellent one to bury cabbage in, as its antiseptic properties preserve them from decay. If the object is to preserve the cabbage for market purposes only, the heads may be buried in the same position in which they grew, or they may be inverted, the stump having no value in itself; but if for seed purposes, they must be buried head up, as whatever injures the stump spoils the whole cabbage for that object. I store between ten and fifty thousand heads annually to raise seed from, and carry them through till planting time with a degree of success varying from a loss for seed purposes of from one-half to thirty-three per cent. of the number buried; but if handled early in Spring, many that would be worthless for seed purposes could be profitably marketed. A few years since I buried a lot with a depth varying from one to four feet, and found, on uncovering them in the Spring, that all had kept and apparently equally well. In the Winter of 1868 excessively cold weather came very early and unexpectedly, before my cabbage plot had received its full covering of litter. The consequence was the frost penetrated so deep that it froze through the heads into the stumps, and when Spring came a large portion of them came out spoilt for seed purposes, though most of them sold readily in the market. A cabbage is rendered worthless for seed when the frost strikes through the stump where it joins the head; and though to the unpracticed eye all may appear right, yet, if the heart of the stump has a water-soaked appearance on being cut into, it will almost uniformly decay just below the head in the course of a few weeks after having been planted out. If there is a probability that the stumps have been frozen through, examine the plot early, and, if it proves so, sell the cabbages for eating purposes, no matter how sound and handsome the heads look; if you delay until time for planting out the cabbage for seed, meanwhile much waste will occur. I once lost heavily in Marblehead Mammoth cabbage by having them buried on a hill-side with a gentle slope. In the course of the Winter they fell over on their sides, which let down the soil from above, and, closing the air chambers between them, brought the huge heads into a mass, and the result was a large proportion of them rotted badly. At another time I lost a whole plot by burying them in soil between ledges of rock, which kept the ground very wet when Spring opened ; the consequence was every cabbage rotted. If the heads are frozen more than two or three leaves deep before they are pitted, they will not come out so handsome in the Spring; but cabbages are very hardy and they readily rally from a little freezing either in the open ground or after they are buried, though it is best when they are frozen in the open ground to let them remain there until the frost comes out before removing them if it can be done without too much risk of freezing still deeper, as they handle better then, for being tougher the leaves are not so easily broken. If the soil is frozen to any depth before the cabbages are removed, the roots will be likely to be injured in the pulling, a matter of no consequence if the cabbages are intended for market, but of some importance if they are for seed raising. Large cabbages are more easily pulled by giving them a little twist ; if for seed purposes this should be avoided, as it injures the stump. A small lot that are to be used within a month can be kept hung up by the stump in the cellar of a dwelling house ; they will keep in this way until Spring, but the outer leaves will dry and turn yellow, the heads shrink some in size, and be apt to lose in quality. Some practice putting clean chopped straw in the bottom of a box or barrel, wetting it, and covering with heads trimmed ready for cooking, adding again wet straw and a layer of heads, so alternating until the barrel or box is filled, after which it is headed up and kept in a cool place, at or a little below the freezing point. No doubt this is an excellent way to preserve a small lot, as it has the two essentials to success, keeping them cool and moist.

Instead of burying them in an upright position, after a deep furrow has been made the cabbages are sometimes laid on their sides two deep, with their roots at the bottom of the furrow, and covered with earth in this position. Where the Winter climate is so mild that a shallow covering will be sufficient protection, this method saves much labor.

HAVING CABBAGE MAKE HEADS IN WINTER.

When a piece of drumhead has been planted very late, (sometimes they are planted on ground broken up after a crop of hay has been taken from it the same season,) there will be a per cent. of the plants when the growing season is over that have not headed. With care almost all of these can be made to head during the Winter. A few years ago I selected my seed heads from a large piece and then sold the first "pick," of what remained at ten cents a head, the second at eight cents, and so down until all were taken for which purchasers were willing to give one cent each. Of course, after such a thorough selling out as this there was not much in the shape of a head left. I now had what remained pulled up and carted away, doubtful whether to feed them to the cows or to set them out to head up during Winter. As they were very healthy plants in the full vigor of growth, having rudimentary heads just gathering in, I determined to set them out. I had a pit dug deep enough to bring the tops of the heads, when the plants were stood upright as they grew, just above the surface of the ground: I then stood the cabbages in without breaking off any of the leaves, keeping the roots well covered with earth, having the plants far enough apart not to crowd each other very much, though so near as to press somewhat together the two outer circles of leaves. They were allowed to remain in this condition until it was cold enough to freeze the ground an inch in thickness, when a covering of coarse hay was thrown over them a couple of inches thick, and, as the cold increased in intensity, this covering was increased to ten or twelve inches in thickness, the additions being made at two or three intervals. In the Spring I uncovered the lot, and found that nearly every plant had headed up. I sold the heads for four cents a pound, and these refuse cabbages averaged me about ten cents a head, which was the price my best heads brought me in the Fall. I have seen thousands of cabbages in one lot, the refuse of several acres that had been planted on sod land broken up the same season a crop of hay had been taken from it, made to head by this course, and sold in the Spring for $1.30 per barrel. When there is a large lot of such cabbages the most economical way to plant them will be in furrows made by the plough. Most of the bedding used in covering them, if it be as coarse as it ought to be to admit as much air as possible while it should not mat down on the cabbages, will, with care in drying, be again available for covering another season, or remain suitable for bedding purposes. These "Winter headed" cabbages, as they are called in the market, are not so solid and have more shrinkage to them than those headed in the open ground ; hence they will not bear transportation as well, neither will they keep as long when exposed to the air. The effect of wintering cabbage by burying in the soil is to make them exceedingly tender for table use.

VARIETIES OF CABBAGE.

If a piece of land is planted with seed grown from two heads of cabbage the product will bear a striking resemblance to the two parent cabbages, with a third variety which will combine the characteristics of these two, yet the resemblance will be somewhat modified at times by a little more manure, a little higher culture, a little better location, and the addition of an individuality that particular vegetables occasionally take upon themselves which we signify by the word "sport." The "sports" when they occur are fixed and perpetuated with remarkable readiness in the cabbage family, as is proved by the great number of varieties in cultivation, the numerous progeny of one ancestor. The catalogues of the English and French seedsmen contain long lists of varieties, many of which (and this is especially true of the early kinds) are either the same variety under a different name or are different "strains" of the same variety produced by the careful selections of prominent market gardeners through a series of years.

Four different seasons I have experimented with foreign and American varieties of cabbage to learn the characteristics of the different kinds, their comparative earliness, size, shape, and hardness of head, length of stump, and such other facts as would prove of value to market gardeners. There is one fact that every careful experimenter soon learns, that one season will not teach all that can be known relative to a variety, and that a number of specimens of each kind must be raised to enable one to make a fair comparison. It is amusing to read the dicta which appear in the agricultural press from those who have made but a single experiment with some vegetable; they proclaim more after a single trial than a cautious experimenter would dare to declare after years spent in careful observation. The year 1869 I raised over sixty varieties of cabbage, importing nearly complete suites of those advertised by the leading English and French seed houses, and collecting the principal kinds raised in this country. I do not propose describing all these in this treatise or their comparative merits; of some of them I have yet something to learn, but I will endeavor to introduce with my description such notes as I think will prove of value to my fellow farmers and market gardeners.

I will here say in general of the class of early cabbages, that most of them have elongated heads between ovoid and conical in form. They appear to lack in this country the sweetness and tenderness that characterize some varieties of our drumhead, and consequently in the North when the drumhead enters the market there is but a limited call for them.

It may be well here to note a fundamental distinction between the drumhead cabbage of England and those of this country. In England the drumhead class are almost wholly raised to feed to stock; I venture the conjecture that this is owing in part, or principally, to the fact that, being raised for cattle, European gardeners have never had the motive and consequently have never developed the full capacity of the drumhead as exampled by the fine varieties raised in this country. The securing of sorts reliable for heading being therefore a matter of secondary consideration, seed is raised from stumps or any refuse heads that may be standing when Spring comes around. For this reason English drumhead cabbage seed is better suited to raise a mass of leaves than heads, and always disappoints our American farmers who buy it because it is cheap with the expectation of raising cabbages for market. English grown drumhead cabbage seed is utterly worthless for use in this country except to raise greens or collards.

The following are foreign varieties that are accepted in this country as standards, and for years have been more or less extensively cultivated: EARLY YORK, LARGE YORK, EARLY OXHEART, LARGE FRENCH OXHEART, EARLY SUGAR LOAF, EARLY WINNIGSTADT. RED DUTCH, RED DRUMHEAD. Of these the Large French Oxheart, Red Drumhead and Early Winnigstadt have had a somewhat recent introduction, the two latter having grown rapidly in popularity. In my experience as a seed dealer, the Sugar Loaf and Oxheart are losing ground in the farming community. The Early Jersey Wakefield having to a large extent replaced them.

Early York. Heads nearly ovoid, pretty hard for an early sort, with few waste leaves surrounding them, which are of a bright green color. Reliable for heading. Stump rather short. Plant two feet by eighteen inches. This cabbage has been cultivated in England over a hundred years. LITTLE PIXIE and CARTER'S SUPERFINE EARLY are with me each of them earlier than Early York, are as reliable for heading, head much harder, and are of better flavor; the first does not grow as large, but the second I think does, and is therefore much preferable to it.

Large York is about a fortnight later than Early York; heads larger, not so long, and more solid; leaves gather closer around the head; stumps short. It is asserted that this variety is less affected by heat than several other kinds, and hence is a good cabbage for the South.

Early Oxheart. Heads nearly egg-shaped, small, hard, few waste leaves, stumps short. A little later than Early York. Have the rows two feet apart, and the plants from sixteen to eighteen inches apart in the row.

Large French Oxheart closely resembles Early Oxheart, but grows to double the size, and is about ten days later; quality usually good. An excellent kind for large early, but, like all others of the oblong headed cabbages, must in this country be marketed before the drumhead class mature or the crop becomes unsalable.

Early Sugar Loaf. Heads shaped much like a loaf of sugar standing on its smaller end, resembling, as Burr well says, a head of Cos lettuce in its shape and in the peculiar clasping of the leaves about the head. Heads rather hard, medium size; early, and tender. It is said not to stand the heat as well as most sorts. Plant in rows two feet apart, and the plants from eighteen to twenty inches in the row.

Early Winnigstadt. (A German cabbage.)

Heads nearly conical in shape having usually a twist of

leaf at the top ; larger than Oxheart, are harder than any of the early oblong heading cabbages; stumps middling short. Matures about ten days later than Early York. The Winnigstadt is remarkably reliable for heading, being not excelled in this respect when the seed has been raised with care, by any cabbage grown. It is a capital sort for Early Market outside our large cities where the very early kinds are not so eagerly craved. It is so reliable for heading that it will often make fine heads where other sorts fail, and I would advise all who have not succeeded in their efforts to grow cabbage to try this before giving up their attempts. It is raised by some for Winter use, and where the drum heads are not so successfully raised I would advise my farmer friends to try the Winnigstadt, as the heads are so hard that they keep without much waste. Have rows two feet apart, and plant twenty inches to two feet apart in the rows.

Red Dutch. Heads nearly conical, medium sized, hard, of a very deep red ; outer leaves numerous, and not so red as the head, being somewhat mixed with green; stump rather long. This cabbage is usually planted too late ; it requires nearly the whole season to mature. It is used for pickling, or cut up fine as a salad served with vinegar and pepper. This is a very tender cabbage, and were it not for its color would be an excellent sort to boil : to those who have a mind to eat it with their eyes shut, this objection will not apply.

Red Drumhead. Like the preceding with the exception that the heads grow round or nearly so, are harder, and of double the size. Care should be taken not to run these cabbages too large, as they will begin to lose in color, which lessens their value for pickling and salad uses. It is very difficult to raise seed from this cabbage in this country. I am acquainted with five trials made in as many different years, two of which I made myself, and all were nearly utter failures, the yield when the hardest heads were selected being at about the rate of two great spoonfuls of seed from every twenty cabbages. French seed growers are more successful, otherwise this seed would have to sell at a far higher figure in the market than any other sort.

The Little Pixie, a recent introduction, has much to recommend it in quality, reliability for heading, and hardness of the head; being earlier than Early York, though somewhat smaller, it is

to be lamented if it does not ultimately sweep away that variety.

Among those that deserve to be heartily welcomed and grow in favor, are the EARLY ULM SAVOY (for engraving and description of which see under head of Savoy,) EARLY VANACK, (a very early conical heading sort,) EARLY NONPAREIL, (another closely allied variety,)—both these latter being among the earliest,—and the ST. DENNIS DRUMHEAD, a late, short-stumped sort, setting a large, round, very solid head, as large but harder than Premium Flat Dutch. The leaves are of a bluish green and thicker than those of most varieties of drumhead. Our brethren in Canada think highly of this cabbage, and if we want to try a new drumhead, I will speak a good word for this one.

Early Schweinfurt or **Schweinfurt Quintal** is a new but excellent early drumhead; the heads range in size from ten to eighteen inches in diameter, varying

with the conditions of cultivation more than any other cabbage I am acquainted with. The heads are flattish round, weigh from three to nine pounds when well grown, are very symmetrical in shape, standing apart from the surrounding leaves. They are not solid, though they have the finished appearance that solidity gives; they are remarkably tender as though blanched, and of very fine flavor. It is among the earliest of drumheads, maturing at about the same time as the Early Winnigstadt. As an early drumhead for the family garden it has no superior; and where the market is near and does not insist that a cabbage head must be hard to be good, it has proved a very profitable market sort.

The following are the standard American varieties of cabbage; EARLY WAKEFIELD, EARLY WYMAN, CRANE'S EARLY, CANNON BALL, EARLY LOW DUTCH, PREMIUM FLAT DUTCH, STONE MASON, LARGE LATE DRUMHEAD, MARBLEHEAD MAMMOTH DRUMHEAD, AMERICAN GREEN GLAZED, FOTTLER'S DRUMHEAD, BERGEN DRUMHEAD, DRUMHEAD SAVOY, and AMERICAN GREEN GLOBE SAVOY. All of these varieties as I have previously stated are but improvements of foreign kinds; but they are so far improved through years of careful selection and cultivation, that as a rule they appear quite distinct from the original kinds when grown side by side with them, and this distinction is more or less recognized in both English and American catalogues by the adjective "American" or "English" being added after varieties bearing the same name.

Early Wakefield, sometimes called **Early Jersey Wakefield.**——
Heads mostly nearly conical in shape but sometimes nearly round, of good size for early, very reliable for heading; stumps short. A very popular early cabbage in the markets of Boston and New York. Plant two and a half feet by two feet.

Early Wyman. This cabbage is named after Capt. Wyman of Cambridge, the originator. Like Early

Wakefield the heads are usually somewhat conical, but sometimes nearly round; in structure they are compact. In earliness it ranks about with the Early Wakefield, and making heads of double the size, it has a high value as an early cabbage. Captain Wyman had entire control of this cabbage until within  the past few years, and consequently has held Boston Market in his own hands to the chagrin of his fellow market gardeners, raising some seasons as many as thirty thousand heads. Have the rows from two to two and a half feet apart, and the plants from twenty to twenty-four inches apart in the row. Crane's Early is a cross between the Wyman and Wakefield, intermediate in size and earliness.

Premium Flat Dutch. A large, late variety; heads either round or flat, on the top, (varying with different strains) rather hard, color bluish-green, leaves around heads rather numerous; towards the close of the season,

the edge of some of the exterior leaves and the top of the heads assume a purple cast. The edges of the exterior leaves and of the two or three that make the outside of the head are quite ruffled so that when grown side by side with Stone Mason, this distinction between the habit of growth of the two varieties is noticable at quite a distance. Stumps short; reliable for heading. Have the rows three feet apart, and the plants from two and a half to three feet apart in the rows. This cabbage is very widely cultivated, and in many respects is an excellent sort to raise for late marketing.

Early Low Dutch. Heads round, medium sized, solid. Outside leaves few in number; stalk thick and short. Medium early, tender and of good quality. Plant two and a half feet by two.

Stone Mason. An improvement on the Mason, which cabbage was selected by Mr. John Mason of Marblehead, from a number of varieties of cabbage that came from a lot of seed purchased and planted as Savoys. Mr. John Stone afterwards improved upon the Mason cabbage, by increasing the size of the heads. Different growers differ in their standard

of a Stone Mason cabbage, in earliness and lateness, and in the size, form and hardness of the head. But all these varieties agree in the characteristics of being very reliable for heading, in having heads, which are large, very hard, very tender, rich and sweet; short stumps, and few waste leaves. The color of the leaves varies from a bluish green to a pea green, and the structure from nearly smooth to much blistered. In their color and blistering some specimens have almost a Savoy cast. The heads of the best varieties of Stone Mason range in weight from six to twenty-five pounds, the difference turning mostly on soil, manure and cultivation.

The Stone Mason is an earlier cabbage than Premium Flat Dutch, has fewer waste leaves, and side by side under high cultivation grows to an equal or larger size, while it makes heads that are decidedly harder and sweeter. These cabbages are equally reliable for heading. I am inclined to the opinion that under poor cultivation the Premium Flat Dutch will do somewhat better than the Stone Mason.

Until the introduction of Fottler's Drumhead it was the standard drumhead cabbage in the markets of Boston and other large cities of the North. Recently this fine cabbage has, in some localities, shown a tendency to rot at the stump before maturing its head. I trust that the trouble is but local and temporary. Have the rows three feet apart, and the plants from two to three feet apart in the row.

Large Late Drumhead. Heads large, round, sometimes flattened at the top, close and firm; loose leaves numerous; stems short; reliable for heading, hardy, and a good keeper. The name "Large Late Drumhead" includes varieties raised by Messrs. Collins & Anderson, Buist, and several other seedsmen in this country, all of which resemble each other in the above characteristics, and differ in but minor points. Have rows three feet apart, and plants from two and a half to three feet apart in the row.

Marblehead Mammoth Drumhead. This is the largest of the cabbage family, having sometimes been grown to weigh over sixty pounds to the plant. It orig-

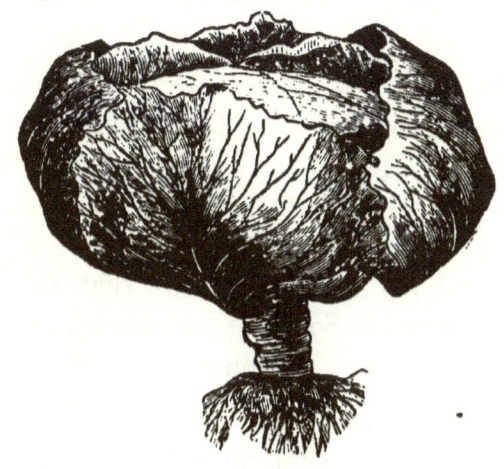

inated in Marblehead, Massachusetts, being produced by Mr. Alley, probably from the Mason, by years of high cultivation and careful selection of seed stock. I introduced this cabbage and the Stone Mason to the general public many years ago, and it has been pretty thoroughly disseminated throughout the United States. Heads varying in shape between hemispherical and spherical, with but few waste leaves surrounding them; size very large, varying from fifteen to twenty inches in diameter, and in some specimens they have grown to the extraordinary dimensions of twenty-four inches. In good soil and with the highest culture this variety has attained an average weight of thirty pounds by the acre. Quality when well grown remarkably sweet and tender, as would be inferred from the rapidity of its growth. Cultivate in rows four feet apart, and allow four feet between the plants in the rows. Sixty tons of this variety have been raised from a single acre.

American Green Glazed. Heads loose though rather large, with a great body of waste leaves surrounding them; quality poor; late; stump long. This cabbage was readily distinguished among all the varieties in my experimental plot by the deep, rich green of the leaves with their bright lustre as though varnished. It is grown somewhat extensively in the South, as it is believed not to be so liable to injury from insects as other varieties. Plant two and a half feet apart each way. I would advise my Southern friends to try the merits of other kinds before adopting this poor affair. I know, through my correspondence, that the Mammoth has done well as far South as Louisiana and Cuba, and the Fottler in many sections of the South has given great satisfaction.

Fottler's Early Drumhead. Several years ago a Boston Seedsman imported a lot of Cabbage seed from Europe, under the name of Early Brunswick Short Stemmed. It proved to be a large heading and very early Drumhead. The heads were from eight to eighteen inches in diameter, near-ly flat, hard, sweet and tender in quality; few waste leaves; stump short. In earliness it was about a fortnight ahead of the Stone Mason. It was so much liked by the Market gardeners that the next season he ordered a larger quantity, but the second importation, though ordered and sent under the same name, proved to be a different and inferior kind, and the same result followed one or two other importations. The two gardeners who received seed of the first importation brought to market a fine large Drumhead, ten days or a fortnight ahead of their fellows. The seed of the true stock was eagerly bought up by the Boston market gardeners, most of it at $5 *an ounce.* After an extensive trial on a large scale by the market farmers around Boston, and by farmers in various parts of the United States, Fottler's Cabbage has given great satisfaction, and has become a universal favorite, and when once known is fast replacing some of the old varieties of Drumhead. Mr. Tillinghast, in his valuable little work on gardening, ranks it emphatically at the head of all the drumhead family, and to those who intend to grow but a single variety, I would heartily recommend the Fottler. Very reliable for heading.

Bergen Drumhead. Heads round, rather flat on the top, solid; leaves stout, thick, and rather numerous; stump short. With me, under same cultivation, it is later than Stone mason. It is tender and of good flavor. A popular sort in many sections, particularly in the markets of New York city. Have the plants three feet apart each way.

Cannon Ball. This cabbage came originally from the Patent office, but as I have been unable to trace its parentage to any foreign country in the course of my experiments with varieties, I think we may as well class it as American by default. The heads are usually spherical, attaining to a diameter of from five to nine inches, with the surrounding leaves gathered rather closely around them; in hardness and relative weight it is not excelled, if equalled, by any other cabbage. Stump short. It delights in the highest cultivation possible. It is about a week later than Early York. In those markets where cabbages are sold by weight, it will pay to grow for market; it is a good cabbage for the family garden.

SAVOY CABBAGE.

The Savoys are the tenderest and richest flavored of cabbages, though not always as sweet as a well grown Stone Mason; nor is a Savoy grown on poor soil or one that has been pinched by drouth as tender as a Stone Mason that has been grown under favoring circumstances; yet it remains as a rule that the Savoy surpasses all other cabbages in tenderness, and in a rich marrow-like flavor. The Savoys are also the hardiest of the cabbage tribe, enduring in the open field a temperature within sixteen degrees of zero without serious injury; and if the heads are not very hard they will continue to withstand repeated changes from freezing to thawing for a couple of months, as far north as the latitude of Boston. A degree of freezing improves them, and it is common in that latitude to let such as are intended for early winter use in the family remain standing in the open ground where they grew, cutting the heads as they are wanted.

As a rule Savoys neither head as readily (the "Improved American Savoy" an exception) nor do the heads grow as large as the drumhead varieties; indeed, most of the kinds in cultivation are so unreliable in these respects as to be utterly worthless for market purposes, and nearly so for the kitchen garden.

The **Drumhead Savoy** sent out by Vilmorin, Andreiex & Co., of France, is not sufficiently distinct from the Green Globe Savoy; it is of a bluish green cast, not so fine in structure, and attains about the same size, but there is not enough of the drumhead in it to make the variety worthy of the name "drumhead." Folsom's American Drumhead Savoy, sometimes called Cambridge Savoy, is much superior, growing to double the size, while it has enough of the Savoy character in it to mark it strongly both for the eye and the palate. One variety in my experimental garden, which I received as Tour's Savoy, (evidently a drumhead variety of the Savoy,) proved to be much like Early Schweinfurt in earliness and style of heading; the heads were very large, but quite loose in structure; I should think it would prove valuable for family use.

It is a fact that does not appear to be generally known that we have among the Savoys some remarkably early sorts which rank with the earliest varieties of cabbage grown. Pancalier and Early Ulm Savoy are earlier than that old standard of earliness, Early York; Pancalier being somewhat earlier than Ulm.

Pancalier is characterized by very coarsely blistered leaves of the darkest green color; the heads usually gather together, being the only exception I know of to the rule that cabbage heads are made up of over-lapping leaves, wrapped closely together. It has a short stump, and with high cultivation is reliable for heading. The leaves nearest the head, though not forming a part of it, are quite tender and may be cooked with the head. Plant fifteen by thirty inches.

Early Ulm Savoy is a few days later than Pancalier, and makes a larger head; the leaves are of a lighter green and not so coarsely blistered; stump short; head round; very reliable for heading. It has a capital characteristic in not being so liable as most varieties to burst the head and push the seed shoot immediately after the head is matured. For  first early I know no cabbages so desirable as these for the kitchen garden.

The **Early Dwarf Savoy** is a desirable variety of second early. The heads are rather flat in shape, and grow to a fair size. Stumps short; reliable for heading.

Improved American Savoy. Everything considered, this is the Savoy "par excellence" for the market garden. It is a true Savoy, the heads grow to a large size, from six to ten inches in diameter, varying, of course, with soil, manure and cultivation. In shape the heads are mostly  globular, occasionally oblong, having but few waste leaves, and grow very solid. Stump short. In reliability for heading it is unsurpassed by any other cabbage.

Golden Savoy differs from other varieties in the color of the head, which rises from the body of light green leaves, of a singular pale yellow color, as though blanched. The stumps are long, and the head rather small, a portion of these growing pointed. It is very late, not worth cultivating except as a curiosity.

Norwegian Savoy. This is a singular half cabbage, half kale—at least, so it has proved under my cultivation. The leaves are long, narrow, tasselated, and somewhat blistered. The whole appearance is very singular and rather ornamental. I have tried this cabbage twice, but have never got beyond the possible promise of a head.

Victoria Savoy, Russian Savoy and **Cape Savoy**, tested in my experimental garden, did not prove desirable either for family use or for market purposes.

Feather Stemmed Savoy. This is a cross between the Savoy and brussels sprouts, having the habit of growth of brussels sprouts.

I will add notes on a few other varieties in my experimental plot:

Large Brunswick Short-Stemmed. (English seed.) Late, long stumped, wild, plenty of leaves, almost no head; bears but a slight resemblance to Fottler's Drumhead.

Early Empress. Cabbages well; heads conical; early.

Robinson's Champion Ox Drumhead. Stump long; heads soft and not very large; wild.

English Winnigstadt. Long stumped; irregular; not to be compared with French stock.

Blenheim. Early; heads mostly conical; of good size.

Shillings Queen. Early; heads conical; stumps long.

Carter's Superfine Early Dwarf. Surpasses in earliness and hardness of head all the early, long headed sorts, Little Pixie, to which it is evidently closely allied, perhaps excepted.

Enfield Market Improved. Most of the heads were flat; rather wild; not to be compared with Fottler.

Kemp's Incomparable. Long headed; heads when mature do not appear to burst as readily as with most of the conical class.

Fielderkraut. Closely resembles Winnigstadt, with larger and longer heads and stump; requires more room than Winnigstadt.

Ramsay's Winter Drumhead. Closely resembles St. Dennis; I think it is the same.

Pomeranian Cabbage. Heads very long; quite large for a conical heading sort; very symmetrical and hard; color yellowish green. It handles well, and I should think would prove a good keeper. Medium early.

Alsacian Drumhead. Stump long; late; wild.

Marbled Bourgogne. Stump long; heads small and hard; color a mixture of green and red.

CABBAGE GREENS.

In the vicinity of our large cities, the market gardeners sow large areas very thickly with cabbage seed, early in the Spring, to raise young plants to be sold as greens. The seed is sown broadcast at the rate of ten pounds and upwards to the acre. Seed of the Savoy cabbage is usually sown for this purpose, which may be sometimes purchased at a discount, owing to some defect in quality or purity that would render it worthless for planting for a crop of heading cabbage.

The young plants are cut off about even with the ground, when four or five inches high, washed, and carried to market in barrels or bushel boxes. The price varies with the state of the market, from 12 cents to $3 a barrel, the average price in Boston market being about a dollar. With the return of Spring most families have some cabbage stumps remaining in the cellar; these can be planted about a foot apart in some handy spot along the edge of the garden, where they will not interfere with the general crop, setting them under ground from a quarter to a half their length, depending on the length of the stumps. They will soon be covered with green shoots, which should be used as greens before the blossom buds show themselves, as they then become too strong to be agreeable. If the spot is rich and has been well dug, the rapidity of growth is surprising; and if the shoots are frequently gathered, many nice messes of greens can be grown from a few stumps. Farmers in Northern Vermont tell me, that if they break off each seed shoot as soon as it shows itself, close home to the stump, nice little heads will push out on almost every stump. In England, where the Winter climate is much milder than that of New England, it is the practice to raise a second crop of heads in this way. I have seen an acre from which a crop of drumhead cabbage had been cut off early in the season, every stump on which had from three to six hard heads, varying from the size of a hen's egg to that of a goose egg; but to get this second growth of heads as much of the stump and leaves should be left as possible, when cutting out the original head. As in the cabbage districts of the North little or no use is made of this prolific after growth, it is worse than useless to suffer the ground to be exhausted by it; the stump should be pulled by the potatoe hoe as soon as the heads are marketed.

When cabbages are planted out for seed, if for any reason the seed shoot fails to push out, and at times when it does push out, fine sprouts for greens will start below the head; when the stock of these sprouts becomes too tough for use, the large leaves may be stripped from them and cooked. I usually break off the tender tops of large sprouts, and then strip off the tenderest of the large leaves below.

CABBAGE FOR STOCK.

No vegetable raised in the temperate zone, Mangold Wurtzel alone excepted, will produce as much food to the acre, both for man and beast, as the cabbage. I have seen acres of the Marblehead Mammoth drumhead which would average thirty pounds to each cabbage, some specimens weighing over sixty pounds. The plants were four feet apart each way, which would give a product of forty tons to the acre; and I have tested a crop of Fottler's that yielded thirty tons of green food to the half acre. Other vegetables are at times raised for cattle feed, such as potatoes, carrots, ruta bagas, mangold wurtzels; a crop of potatoes yielding four hundred bushels to the acre at sixty pounds the bushel would weigh twelve tons; a crop of carrots yielding twelve hundred bush. els to the acre would weigh thirty tons; and ruta bagas sometimes yield thirty tons, and mangolds as high as seventy tons to the acre. I have set all these crops at a high capacity for fodder purposes; the same favoring conditions of soil, manure, and cultivation that would produce four hundred bushels of potatoes, twelve hundred bushels of carrots, and thirty-five tons of ruta baga turnips, would give a crop of forty tons of the largest variety of drumhead cabbage. If we now consider the comparative merits of these crops for nutriment, we find that the cabbage excels them all in this department also. The potato abounds in starch, the mangold and carrot are largely composed of water, while the cabbage abounds in rich, nitrogenous food, ranking in nutriment almost side by side with the flesh of animals.

When cabbage is kept for stock feed later than the first severe frost, if the quantity is large there is considerable waste even with the best of care. The loose leaves should be fed first, and the heads kept on the stump in a cool place, not more than two or three deep, at as near the freezing point as possible. If it has been necessary to cut the heads from the stumps, they may be piled, after the weather has set in decidedly cold, conveniently near the barn, and kept covered with a foot of straw or old litter. As long as a cabbage is kept frozen there is no waste to it; but if it be allowed to freeze and thaw two or three times, it will soon rot with an awful stench. On the other hand, if it is kept in too warm and dry a place, the outer leaves will dry, turning yellow, and the whole head lose in weight,—if it be not very hard, shriveling, and if hard, shrinking. If they are kept in too warm and wet a place, the heads will decay fast, in a black, soft rot. The best way to preserve cabbages for stock into the winter is to place them in trenches a few inches below the surface, and there cover with from a foot to two feet of coarse hay or straw, the depth depending on the coldness of the locality. When the ground has been frozen too hard to open with a plough or spade, I have kept them until Spring by piling them loosely, hay-stack shape, about four feet high, letting the frost strike through them, and afterwards covering with a couple of feet of eel grass; straw or coarse hay would doubtless do as well.

I have treated of cabbage thus far when grown specially for stock; in every piece of cabbage handled for market purposes, there is a large proportion of waste suitable for stock feed, which includes the outside leaves and such heads as have not hardened-up sufficiently for market. On walking over a piece of three or four acres one Fall, just after my cabbages for seed stock had been taken off, I noted that the refuse leaves that were stripped from the heads before pulling were so abundant that they nearly covered the ground. If leaves so stripped remain exposed to frost, they soon spoil; or, if earlier in the season they are exposed to the sun, they soon become yellow, dry, and of but little value. They can be rapidly collected with a hay fork and carted, if there be but a few, into the barn; should there be a large quantity, dump them within a convenient distance of the barn or feeding ground, but not where the cattle can trample them, and spread them so that they shall be but a few inches in depth. If piled in heaps they will quickly heat; but even then, if not too much decayed, cattle will eat them with avidity.

If cabbage is fed to cows in milk without some care, it will be apt to give the milk a strong cabbage flavor; all the feed for the day should be given early in the morning. Beginning with a small quantity, and gradually increasing it, the dairyman will soon learn his limits. The effect of a liberal feed to milk stock is to increase the flow of milk, under some circumstances more than two fold. Avoid feeding to any extent while the leaves are frozen.

An English writer says; "The cabbage comes into use when other things begin to fail, and it is by far the best succulent vegetable for milking cows—keeping up the yield of milk, and preserving better than any other food some portion of the quality which cheese loses when the cows quit their natural pasturage. Cows fed on cabbages are always quiet and satisfied, while on turnips they often scour and are restless. When frosted they are liable to produce hoven unless kept in a warm shed to thaw before being used; fifty-six pounds given, at two meals, are as much as a large cow should have in a day. Frequent cases of abortion are caused by an over supply of green food. Cabbages are excellent for young animals, keeping them in health, and preventing 'black leg.' A calf of seven months may have twenty pounds a day."

RAISING CABBAGE SEED.

Cabbage seed in England, particularly of the drumhead sorts is mostly raised from stumps, or from the refuse that remains after all that is salable has been disposed of. The agent of one of the largest English seed houses, a few years since, laughed at my " wastefulness" as he termed it, in raising seed from solid heads. In this country cabbage seed is mostly raised from soft, half-formed heads, which are grown as a late crop, few, if any of them, being hard enough to be of any value in the market. Seedsmen practice selecting a few fine hard heads from which to raise their seed stock. It has been my practice to grow seed from none but extra fine heads, better than the average of those carried to market. I do this on the theory that no cabbage can be too good for a seed head, if the design is to keep the stock first class. Perhaps such strictness may not be necessary, but I had rather err in setting out too good heads than too poor ones; besides, the great hardness obtained by the heads of the Stone Mason makes it possible, at least, that I am right. Cabbage raised from seed grown from stumps are apt to be unreliable for heading and to grow long-stumped, though under unfavorable conditions long stumped and poor headed cabbage may grow from the best of seed. To have the best of seed all shoots that start below the head should be broken off. The shoots should be protected from the wind by being tied to stakes, and scarecrows should be set up, or some like precaution be taken to keep away the little seed birds that begin to crack the pods as soon as they commence to ripen. A plaster cat is a very good scarecrow to frighten away birds from seed and small fruits, if its location is changed every few days.

I find that the pods of cabbage seed grown South are tough, and not brittle like those grown North, and hence that they are injured but little, if any, by seed birds. When the seed pods have passed what seedsmen call their "red" stage, they begin to harden; as soon as a third of them are brown the entire stalk may be cut and hung up in a dry, airy place, for a few days, when the seed will be ready for rubbing or threshing out. Different varieties should be raised far apart to insure purity; and cabbage seed had better not be raised in the vicinity of turnip seed. There is some difference of opinion as to the effect of growing these near each other; where the two vegetables blossom at the same time, I should fear an admixture. When the care requisite to select good seed stock and the trouble of keeping it over winter, planting it in isolated locations, protecting it from wind and weather, guarding it from injury from birds and other enemies, gathering it, cleaning it, are all considered, few men will find that they can afford to raise their own seed, provided they can buy it from reliable seedsmen.

COOKING CABBAGE, SOUR KROUT, &c.

Cabbage when boiled with salt pork, as it is mostly used, is the food for strong and healthy digestive powers; but when eaten in its raw state served with vinegar and pepper it is considered one of the most easily digested articles of diet. In the process of cooking, even with the greatest care, a large portion of the sweetness is lost. The length of time required to cook cabbage by boiling varies with the quality, those of the best quality requiring about twenty minutes, while others require an hour. In cooking put it into boiling water in which a little salt or soda has been sprinkled which will tend to preserve the natural green color. It will be well to change the water once. The peculiar aroma given out by cabbage when cooking is thought to depend somewhat on the manner in which it is grown; those having been raised with the least rank manure having the least. I think this one of the whims of the community.

To *Pickle*, select hard heads, quarter them, soak in salt and water four or five days, then drain and treat as for other pickles, with vinegar spiced to suit.

For *Cold Slaw*, select hard heads, halve and then slice up these halves exceedingly fine. Lay these in a deep dish, and pour over vinegar that has been raised to the boiling point in which has been mixed a little pepper and salt.

Sour Krout. Take large, hard headed drumheads, halve and cut very fine, then pack in a clean, tight barrel, beginning with a sprinkling of salt and following with a layer of cabbage, and thus alternating until the barrel is filled. Now compact the mass as much as possible by pounding, after which put on a well fitting cover resting on the cabbage, and lay heavy weights or a stone on this. When fermented it is ready for use. To prepare for the table fry in butter or fat.

The outer green leaves of cabbages are sometimes used to line a brass or copper kettle in which pickles are made, in the belief that the vinegar extracts the coloring substance (chlorophyl) in the leaves, and the cucumbers absorbing this acquire a rich green color. Be not deceived by this transparent cheat, O simple housewife! the coloring matter comes almost wholly from the copper or brass behind those leaves ; and, instead of an innocent vegetable pigment, your green cucumbers are dyed with the poisonous carbonate of copper.

CABBAGES UNDER GLASS.

The very early cabbages usually bringing very high prices, the enterprising market gardener either winters young plants under glass or starts them there, planting the seed under its protecting shelter long before the cold of Winter is passed. When the design is to winter over Fall grown plants, the seed are planted in the open ground about the middle of September and at about the last of October they are ready to go into the cold frames, as such are called that depend wholly on the sun for heat. Select those having short stumps and transplant into the frames, about an inch and a half by two inches apart, shading them with a straw mat or the like for a few days, after which let them remain without any glass over them until the frost is severe enough to begin to freeze the ground, then place over the sashes, but bear in mind that the object is not to promote growth, but as nearly as possible to keep them in a dormant state, to keep them so cold that they will not grow, and just sufficiently protected to prevent injury from freezing. With this object in view the sashes must be raised whenever the temperature is above freezing, and this process will so harden the plants that they will receive no serious injury though the ground under the sash should freeze two inches deep; cabbage plants will stand a temperature of fifteen to twenty degrees below the freezing point. A covering of snow on the sash will do no harm, if it does not last longer than a week or ten days, in which case it must be removed. There is some danger to be feared from ground mice, who, when everything else is locked up by the frost will instinctively take to the sash, and there cause much destruction among the plants unless these are occasionally examined. When March opens remove the sash when the temperature will allow, replacing it when the weather is unseasonably cold, particularly at night. The plants may be brought still farther forward by transferring them from the hot bed when two or three inches high to cold frames, having first somewhat hardened them. When so transferred plant them about an inch apart, and shield from the sun for two or three days. After this they may be treated as in cold frames. The transfer tends to keep their stock, increases the fibrous roots and makes the plants hardier. As the month advances it may be left entirely off, and about the first of April the plants may be set out in the open field, pressing fine earth firmly about the roots.

When cabbages are raised in hot beds the seed in the latitude of Boston should be planted the first of March; in that of New York about a fortnight earlier. When two or three inches high, which will be in three or four weeks, they should be thinned to about four or less to an inch in the row. They should now be well hardened by partly drawing off the sashes in the warm part of the day, and covering at night; as the season advances remove the sashes entirely by day, covering only at night. By about the middle of April the plants will be ready for the open ground.

When raised in cold frames in the Spring, the seed should be planted about the first of April, mats being used to retain by night the solar heat accumulated during the day. As the season advances the same process of hardening will be necessary as with those raised in hot beds.

COLD FRAME AND HOT BED.

To carry on hot beds on a large scale successfully is almost an art in itself—and for fuller details I will refer my readers to works on gardening. Early plants in a small way may be raised in flower pots or boxes in a warm kitchen window. It is best if practicable, to have but one plant in each pot that they may grow short and stocky. If the seed are not planted earlier than April for out of door cultivation a cold frame will answer.

For a Cold Frame select the locality in the Fall, choosing a warm location on a southern slope, protected by a fence or building on the north and northwest. Set posts in the ground, nail two boards to these parallel to each other, one about a foot in height, and the other towards the south about four inches narrower; this will give the sashes resting on them the right slope to shed the rain and receive as much heat as possible from the sun. Have these boards at a distance apart equal to the length of the sash, which may be any common window sash for a small bed, while three and a half feet is the length of a common gardener's sash. If common window sash is used cut channels in the cross bars to let the water run off. Dig the ground thoroughly (it is best to cover it in the Fall with litter to keep the frost out) and rake out all stones or clods; then slide in the sash and let it remain closed three or four days that the soil may be warmed by the sun's rays. The two end boards and the bottom board should rise as high as the sash to prevent the heat escaping, and the bottom board of a small frame should have a strip nailed inside to rest the sash on. Next rake in thoroughly guano or phosphate or finely pulverized hen manure, and plant in rows four to six inches apart. As the season advances raise the sashes an inch or two in the middle of the day and water freely at evening with water that is nearly of the temperature of the earth in the frame. As the heat of the season increases whitewash the glass and keep them more and more open until just before the plants are set in open ground, then allow the glass to remain entirely off both day and night unless there should be a cold rain. This will harden them so that they will not be apt to be injured by the cabbage beetle, as well as chilled and put back by the change. Should the plants be getting too large before the season for transplanting, they should be checked by drawing a sharp knife within a couple of inches of the stalk. If it is desirable to check their growth or harden them, transplant into another cold frame, allowing each plant double the distance it before occupied.

The structure and management of a Hot Bed is much the same as that of a Cold Frame, with the exception that the sashes are usually longer and the back and front somewhat higher; being started earlier the requisite temperature has to be kept up by artificial means, fermenting manure being relied upon for the purpose, and the loss of this heat has to be checked more carefully by straw matting, and in the far North by shutters also. In constructing it horse manure with plenty of litter and about a quarter its bulk in leaves, if attainable, all having been well mixed together, is thrown into a pile, and left for a few days until steam escapes, when the mass is again thrown over and left for two or three days more, after which it is thrown into the pit (or it may be placed directly on the surface) which is lined with boards, from eighteen inches to two feet in depth, when it is beaten down with a fork and trodden well together. The sashes are now put on and kept there until heat is developed. The first intense heat must be allowed to pass off, which will be in about three days after the high temperature is reached. Now throw on six or eight inches of fine soil in which mix well rotted manure free from all straw, or rake in thoroughly Superphosphate or Guano, at the rate of two thousand pounds to the acre and plant the seed as in Cold Frame. Harden the plants as directed in preceding paragraph.

CAULIFLOWER, BROCCOLI, BRUSSELS SPROUTS, KALE, AND SEA KALE.

My treatise on the cabbage would hardly be complete without some allusion to such prominent members of the Brassica family as the cauliflower, broccoli, brussels sprouts, and kale. These in the selection and preparation of the soil, manure, and cultivation require for the most part the same treatment as cabbage. In Europe there has been far more progress made in the cultivation and use of these vegetables than with us in America; and I am indebted to the work of McIntosh for many of my ideas in this section. The Broccoli are closely allied to the cauliflower, the white varieties bearing so close a resemblance that one of them, the Walcheren, is by some classed indiscriminately with each. The chief distinction between the two is in hardiness, the broccoli being much the hardier.

The **Cauliflower** require the same distance apart in the rows and between the plants as cabbage, the early and late varieties corresponding in this respect with the early and late varieties of cabbage. To perfect them the very highest cultivation possible is required; give them strong, deep soil, very thoroughly worked: use liquid manure freely and water abundantly. A fine cauliflower is the pet achievement of the market gardener. The great aim is not to produce size only, " but the fine, white, creamy color, compactness, and what is technically called curdy appearance, from its resemblance to the curd of milk in its preparation for cheese. When the flower begins to open, or when it is of a warty or frost-like appearance, it is less esteemed. It should not be cut in summer above a day before it is used." The cauliflower is served with milk and butter, or it may become a component of soups, or be used as a pickle. Many of the varieties given in catalogues are but synonyms of, and very closely resemble each other. Among the most desirable for cultivation are EXTRA EARLY ERFURT, HALF EARLY PARIS or DEMI DUR, (this is the kind usually sold in this country as Early Paris, the true variety making so small a head as to be comparatively worthless here), WALCHEREN, LARGE ASIATIC, NONPAREIL, LENORMAND.

The leaves of the Extra Early Erfurt growing close to the head permit its being planted nearer than any other early sort. I have grown this variety with heads fifteen inches in diameter.

Of **Broccoli** over forty varieties are named in foreign catalogues, of which WALCHEREN is one of the very best. KNIGHT'S PROTECTING is an exceedingly hardy dwarf sort. As a rule the white varieties are preferred to the purple kinds. Plant and treat as cauliflower.

Of **Brussels Sprouts** (or bud-bearing cabbage) there are but two varieties, the dwarf and the tall; the tall kind produces more buds, while the dwarf is the hardier. The "sprouts" form on the stalks, and are miniature heads of cabbage from the size of a pea to that of a pigeon's egg. They are raised to but a limited extent in this country, but in Europe they are grown on a large scale. The sprouts may be cooked and served like cabbage, though oftentimes they are treated more as a delicacy and served with butter or some rich sauce. The FEATHER STEM SAVOY and DALMENY SPROUTS are considered as hybrids, the one between the brussels sprouts and Savoy, the other between it and Drumhead Savoy. The soil for brussels sprouts should not be so rich as for cabbage, as the object is to grow them small and solid. Give the same distance apart as for early cabbage, and the same manner of cultivation. Break off the leaves at the sides a few at a time when the sprouts begin to form and when they are ready to use cut them off with a sharp knife.

Kale. Sea kale or sea cabbage is a native of the sea coast of England, growing in the sand and pebbles of the sea shore. It is a perennial, perfectly hardy, withstanding the coldest winters of New England. The blossoms, though bearing a general resemblance to those of other members of the cabbage family, are yet quite unique in appearance, and I think worthy of a place in the flower garden. It is propagated both .by seed and by cuttings of the roots, having the rows three feet apart, and the plants three feet apart in the rows. It is difficult to get the seeds to vegetate. Plant seed in April and May. The ground should be richly manured and deeply and thoroughly worked. It is blanched before using. In cooking it requires to be very thoroughly boiled, after which it is served up in melted butter and toasted bread. The sea kale is highly prized in England, but thus far its cultivation in this country has been very limited.

The **Borecole** or common kale is of the cabbage family, but is characterized by not heading like the cabbage or producing eatable flowers like the cauliflower and broccoli. The varieties are very numerous, some of them growing very large and coarse, suitable only as food for stock ; others are exceedingly finely curled, and excellent for table use ; while others in their color and structure are highly ornamental. They are annual, biennial, and perennial. They do not require so strong a soil or such high manuring as other varieties of the cabbage family.

The varieties are almost endless : some of the best in cultivation for table use are DWARF GREEN CURLED or GERMAN GREENS, TALL GREEN CURLED, PURPLE BORECOLE, and the variegated kales. The crown of the plant is used as greens, or as an ingredient in soups. The kales are very hardy, and the dwarf varieties with but little protection can be kept in the North well into the

Winter in the open ground. Plant and cultivate like Savoy cabbage.

The variegated sorts with their fine curled leaves of a rich purple, green, red, white, or yellow color, are very pleasing in their effects, and form a striking and attractive feature when planted in clumps in the flower garden-particularly is this so because their extreme hardiness leaves them in full vigor after the cold has destroyed all other plants,—some of the richest colors are developed along the veins of the uppermost leaves after the plant has nearly finished its growth for the season. The JERSEY COW KALE grows to from three to six feet in height and yields a great body of green food for stock; have the rows about three feet apart, and the plants two to three feet distant in the rows. In several instances my customers have written me that this kale raised for stock feed has given them great satisfaction.

The THOUSAND HEADED KALE is a tall variety sending out numerous side shoots, whence the name.

ONION RAISING;

WHAT KINDS TO RAISE,

AND

THE WAY TO RAISE THEM.

EIGHTEENTH EDITION (REVISED).

JAMES J. H. GREGORY,
SEED GROWER, MARBLEHEAD, MASS.

MARBLEHEAD:
MESSENGER STEAM PRINTING-HOUSE.
1882.

Entered according to act of Congress, in the year 1864,
By JAMES J. H. GREGORY,
At the Clerk's Office of the District Court, District of Massachusetts.

ONION RAISING.

WHY I WRITE THIS TREATISE.

In common with my fellow-seedsmen I frequently receive letters from my farmer friends, in different parts of the United States, asking for information on Onion Raising. It is impossible in a letter sheet to give sufficient minuteness of detail; I therefore send out this little treatise, in which I have endeavored to cover very minutely the whole ground of inquiry. I trust that it will prove acceptable.

SELECTING THE SOIL.

Onions are an exception to the general rule,—they thrive best on old ground, with the exception of an increased liability to injury from rust or smut. I recently examined an acre of land which had been planted continuously with onions for three generations without perceptible decrease in the quantity or quality of the crop.

Onions are sometimes successfully raised by plowing up old pasture land in September, thoroughly harrowing it before frost sets in, and in the spring working in fine manure very thoroughly with the harrow and cultivator. The result of such planting is to get a crop very free from weeds, with onions usually coarse, and more or less of scallions.

Onions can be raised on a variety of soils, but yield the most satisfactory returns on a sandy loam, a gravelly soil, or, to state a general rule, on those soils which are light in structure. As onions are brought on the heavier soils, the first effect will be a deterioration in their appearance, the outer skin of the yellow varieties, losing its fine, clear, translucent yellow, and becoming thicker, duller, and less attractive in appearance. If planted on a wet or very heavy soil, the crop will mature late, if it matures at all, giving a large proportion of that dread of the onion grower, scallions, or "scullions" as farmers term them, meaning those whose growth runs mostly to the neck, forming little or no bulb or bottom. With plenty of manure onions will thrive well on soil that is very gravelly. I have seen very large crops grown on Marblehead Neck, on land so stony, that, after a rain, on an area of many square yards not a particle of soil could be seen, nothing but small angular fragments of porphyry, with thrifty onions springing as it were out of the very rocks. Let it be understood, however, that this soil was not of a leachy nature, but rested on a hard-pan bottom. The area of land selected should be free of all large stones, as such interfere seriously with the straightness of the rows, the planting, hoeing and general cultivation of so small-sized a product. Ultimately, good cultivators clear their onion grounds of large loose rocks by blasting or sinking them; obviously, the sooner this is done, the larger are the returns received from such judicious investments. The land should be laid out in as nearly a square as practicable, as this facilitates estimates of manure, seed, and crops, gives greater regularity to the work, and economizes in the cultivation of a crop which requires a great deal of passing over. To protect the crop from the washing of heavy showers, the land should be level or very nearly so, otherwise a rush of water will bare the roots of some, and heap the earth around the necks of others, to the injury of each.

PREPARING THE SOIL.

Don't plant a weedy soil to onions, or land which abounds in witch, or couch grass; if you do, you will repent it on your hands and knees all summer long, for such soil will usually require two more weedings than that on which weeds have not been allowed to ripen their seed. To have to keep down witch grass with your fingers in an onion bed is a miserable business, tearing up the onions and your patience at the same time; better delay a year, and meanwhile clear the land thoroughly by a diligent use of the cultivator and hoe, finishing in the fall by throwing the land into ridges that the freezings and thawings of winter may act destructively on the roots of the witch grass. Should any scattered shoot of this grass show itself in the spring, let the roots be carefully removed with a fork or spade before the land is plowed.

When onions are planted on land full of the seed of weeds it is well, if the season is an early one, to give sufficient time for the first crop of weed seed to start before planting the onions.

In the Eastern States it is found, as a general rule, that success with the first crop of onions is affected by the crop which grew in the land the previous year, and that onions will follow carrots better than any other crop; next to carrots, corn and potatoes are ranked as good preparers of the ground, while to succeed well with onions where cabbage or beets were raised the previous year is comparatively rare. Were there no other reason, the clean tilth which carrots insure makes it an excellent crop to precede onions. In the fertile lands of the west, the method of procedure is briefly this: Land on which grows the bush-hazel is selected, if accessible, the bushes cut down and the turf surface but little more than pared in spring with the plow. In this condition it is usually allowed to remain a season, exposed to the drying effects of the sun, when it is most thoroughly harrowed and raked, and all the numerous roots and waste are burnt, the land plowed to a moderate depth, and the seed sown either broadcast or in drills. Should the early part of the season prove very wet, the crop sowed broadcast is at times smothered under a rapid growth of weeds, while with a favoring season as high as 800 bushels to the acre have been harvested.

After the harvesting of the crop which is to precede onions, let the land have a fall plowing, and be thrown up into ridges, which will not only help destroy noxious weeds and witch grass as above stated, but will leave the land light, in a condition to be worked successfully early in the spring—a great desideratum for a crop that usually requires the entire season to mature it.

THE MANURE.

Onions require the very best of manure, in the most tempting condition, and plenty of it at that. Peruvian guano, fish guano, pig manure, barn manure, night-soil, kelp, muscle mud, superphosphate of lime, wood ashes, and muck are, either alone or in compost, all excellent food for the onion. Old ground, to maintain it in first-rate condition, should receive from six to eight cords of manure to the acre; while new onion ground, to get it in first-rate condition, should receive from eight to ten cords of manure. When Peruvian guano was held at about sixty dollars per ton, experienced farmers believed that no purchased manure paid as well as this on old beds, provided two applications were made, one of about 500 lbs. to the acre, to be raked in at the time of planting, and the other of like amount to be applied broadcast when the onions were about half grown. Those who used but one application at the time of sowing were apt to see surprising effects in a fine growth up to the period of half maturity of the crop, and an equally surprisingly effect in but little growth from this time through the remainder of the season. Those who have used guano freely on their onion lands in the vicinity of Philadelphia assert that one singular result is, that, after applying it for three years in succession, the seed onions for the most part fail to sprout in such soil, and when seed is planted it makes but little growth after vegetating. As far as I have observed, superphosphate of lime used as a manure for a series of years is apt to give the first of these results. Pig manure is held in high esteem by many successful growers of onions in southern New England. Fish guano applied at the rate of a ton to the acre has given very fine crops.

In the vicinity of large towns, where night-soil can be readily obtained, no more efficient manure can be applied than a compost of this and muck that has been exposed to a winter's frost, or good loam, in the proportion of three parts muck or loam to one part night-soil. If with this compost barn manure and sea manure are mixed, so much the better; for it is a rule for this as for other crops that a combination of manures in an arithmetical ratio will produce results approaching a geometrical ratio. To make a compost of loam or muck and night-soil, select a spot very near the piece to be planted, and cover the ground with either to the depth of a foot or eighteen inches; then raise a bank of the same material surrounding this floor to the height of three or four feet, with a thickness of from four to six feet.

The carts containing night-soil are backed up against this receptacle, and the door being unscrewed, the contents shoot out. If barn manure is used, it usually forms part of the sides of the receptacle. During the winter the frosts act on the heap to the further sweetness and disintegrating of it, and towards spring the mass is pitched most thoroughly over, being mixed and made as fine as possible,—sand when obtainable having been either previously, or being subsequently liberally mixed with it, which so "cuts" or separates it that it remains light and fine. After an interval of about a fortnight, allowing time for fermentation, the heap is again pitched over for fining and mixing, and, occasionally, three mixings are made. It would be well for tourists to avoid the onion districts at this season of the year, as a little experience will amply satisfy them.

These composts should not be made on the ground where the onions are to be planted, for neither onions nor any other crop will grow on such spots the same season.

Where superphosphate of lime is used, it is best to make two applications, as with guano. The results of the use of superphosphates are not always satisfactory, but I have seen eight hundred pounds applied to the acre produce as good results as seven cords of rich compost applied side by side.

Muscle mud obtained from the sea-coast is rarely used alone, though large crops are sometimes raised on old onion ground by the application of this alone, at the rate of eight cords to the acre. It appears to give the best results a few miles inland. The strength and consequent value of this manure varies considerably; and here let me add that the value of all animal manures will be found to vary greatly; other things being equal, the higher feeding the animal receives, the better the manure.

Wood ashes are generally used in connection with other manures at the rate of about 200 bushels to the acre. Wood ashes should never be *combined* with other manures, as it will set the ammonia free, and thus deteriorate their quality. Use ashes either by scattering it on the surface at the time of planting, or when the crop is about half grown.

In the vicinity of large towns, of all manures obtained outside the barnyard, night-soil is the cheapest. The first farmer who used it in this locality, comparatively but a few years ago, was universally jeered at by his comrades, but now nearly all of our annual crop of 50,000 bushels is fed principally on this manure.

The effect of kelp, (by this I mean the sea-manure which is thrown up by the storms on very bold shores,) when used as the principal manure, is to give a coarse onion, and a late crop; so late as oftentimes to be in quite a green state at the close of the season, requiring extra labor and care to get it in market condition. In seasons of great drought, however, kelp serves an excellent end, in so retarding the crop that it is not prematurely ripened. In the excessively dry season of 1864, crops along the sea-coast manured with kelp, in many instances yielded double those manured with barnyard and other manures.

The manure is managed most conveniently by dropping it on the land in quite small heaps, at regular intervals, at convenient distance for spreading. I close this paragraph on manures by emphasizing the utility of *a thorough fining of it*.

PLOWING.

The farmer who brings up the sub-soil on his onion bed, will find he has made a mistake. Onions do not require deep plowing; four or five inches is sufficient depth to insure a good crop. One of the finest pieces I ever saw was managed by carting on the manure in the fall, and simply giving it a thorough working into the soil with an ordinary one-horse cultivator in the spring, after which the land was raked and planted, no plow or any implement other than the cultivator having been used. In this instance the soil was naturally quite light. In the west, the ground having been plowed in the fall, it frequently receives only a cultivating or harrowing in the spring.

As the great object is to get the land in a thoroughly fine condition, to facilitate the covering of the seed with fine earth, to leave the soil light that there may be a vigorous growth of the plants, and to leave the land in good working condition for after culture, no labor should be spared to attain this end. On most soils the ground should be plowed, cross-plowed, and thoroughly cultivated. If, from the backwardness of the spring and the consequent wet state of the land, the soil should still be lumpy, it should be thoroughly rolled before raking for planting, and it may be well to brush-harrow it. As onions grown from the seed usually require the whole season to ripen, the onion grower breaks ground first of all in his onion bed, springing to this as early in the season as is possible to work the land into a light and fine condition.

THE SEED.

In some localities three pounds of seed was thought sufficient to an acre; afterward this was increased to three and a half, and then to four; and now, when raised for tracing, five and six pounds are sometimes planted. As a general rule, three and one-half pounds will be found sufficient for an acre; and when land is very heavily manured, four pounds may be planted with profit. Land that is planted to onions the first time requires more seed than old land. If it is designed to pull the onions when small for bunching for the early market, then seven or eight pounds of seed will be required for an acre. If the intent is to raise the very small onions known as "setts," which are stored over winter to be planted in the spring to produce early onions, than a much larger quantity will be required. On old beds where rust abounds, I have known sixteen pounds of seed sown to the acre. Of course it is of the first importance that the seed should be reliable.

Compared with the average return of the crop, the cost of good seed for planting an acre of land to onions, even at the highest prices, is not to be considered a moment beside the acceptance of doubtful seed *even as a gift;* yet every onion-growing community has had its stories to tell of cultivators who have thrown away their time, labor and manure, by purchasing doubtful seed at a little lower figure than that at which reliable seed could be procured. New seed will sometimes fail to vegetate if planted a little too deep, or if snow falls and remains on the ground after planting, or a rain falls after raking and just before planting, though part of the same piece planted but an hour before may come finely.

The usual test for good seed, that is, seed that will vegetate, is the sinking of it; that which will sink being considered reliable, and that which floats being considered worthless. This will answer as a general rule, but it is not wholly reliable. Seed that will sink will not always vegetate, while seed that will float, under some circumstances, will vegetate. Any farmer who tests his seed by the sinking process will find that some of that which floats will vegetate, while no farmer is safe in planting seed that is two years old, though it will sink. Some farmers ascertained this latter fact to their great loss during the spring of 1864. If the season is an average one, such seed as is two years old can be relied upon if it has weight sufficient to sink it; but such a season as the spring of 1864, being unusually wet, much of the two-years seed that was sown, though sown by farmers who had themselves raised it, failed to vegetate.

The lesson to be learned by such unfortunate result is, that it is never perfectly safe to sow seed that is two years old, and that the only way a prudent cultivator will use it will be when mixed with a large proportion of fresh seed. There are two special risks incidental to the sinking test; first, the danger that the seed will not be thoroughly dried, as onion seed when containing sufficient moisture to cause it to sprout if stored in bulk, appears dry to the eye; again, the vitality of onion seed is very apt to be hurt by the drying of it, particularly so, as it is usually deferred until just previous to planting, when matters are greatly hurried, (as the risk of injury through this process is considered too great to permit it to be sunk earlier in the season,) and then it is likely to be exposed too near the kitchen stove. Seed thoroughly winnowed by the wind, on a large sheet spread on some open spot, free from all eddies will be found to give a quality very nearly or quite as free from light seed as the sinking process. As the objection just presented does not lie against this process, it is decidedly preferable. The only reliable test for the vitality of any variety of seed is that which includes all the usual conditions of growth. Testing by planting in a hot-house or in a box in a common house, is not fully reliable, because the seed are not surrounded by the conditions of natural growth,—they then have a temperature very mild, and very nearly constant, with no excess of moisture or dryness,—whereas the natural condition of vegetation includes the very varying temperature of early spring, usually a great excess of moisture and a low degree of heat, all of which causes, either single or combined in their effects, draw largely on the vital power of the seed. Hence, seed that under the favoring influences of the hot-house or kitchen may vegetate, may not have sufficient vitality to overcome the excessive cold or moisture of the garden. The result, therefore, of the usual experimental tests can be relied upon as giving only an approximation to the truth.

Among these approximate tests is the simple one of partially filling a tumbler with cotton-wool, pouring in a little water, not sufficient to cover the cotton, then sprinkling a certain number of seed on the cotton, covering it with a little additional cotton to keep the moisture in. Another simple test is to sprinkle the seed to be tested on a moist woolen cloth, fold the cloth together, and put it in a place moderately warm. The proportion of seed that is good will be known by the proportion that sprout. Experienced eyes can learn something by the appearance and feel of the seed. Old seed require several days longer to vegetate than new.

WHAT KIND OF ONIONS TO PLANT.

Foreign catalogues describe a score and more varieties of onion which are raised in Europe, but as far as experiments have been made with them in this country, it has been found that European-grown onion seed cannot be relied upon to give as good bulbs as American-grown of the same varieties; while many sorts are not adapted to our climate. A measure of the dubious quality of this foreign seed is well indicated by the lower price at which it is generally catalogued.

Of those grown from seed, the Large Red, Yellow, and White are the three standard varieties in the United States.

The Large Red is commonly known as the Wethersfield onion, it having been extensively cultivated in that locality at an early day in onion culture. This is commonly divided into four varieties, viz:

LATE LARGE RED (see illustration) is a very large, thick, late onion, attaining a diameter of from three to six inches, and on the fertile prairies of the west, not unfrequently eight inches.

SECOND EARLY, which differs only in size and time of ripening; being rather flatter than the large sort, not so large, and coming to maturity earlier.

THE EARLY FLAT RED is still flatter in form, smaller in size, rather light-colored, and matures earliest of the three sorts; as early as the last of July.

There is also a fine GLOBE variety of EARLY RED onion

(see illustration) in cultivation, which comes to maturity about a week earlier than the Danvers Early, is of good size and flavor, and in color usually of a very bright, handsome red. The seed of this variety is much sought after by onion-growers, but it is difficult to procure in a pure state.

There is a very handsome late variety of onion known as Southport Red Globe, which originated in Southport, Connecticut. It is quite late and therefore not safe to plant north of Connecticut.

There are four varieties of the Yellow onion in cultivation, of which the Yellow Flat, called also Yellow Dutch, and Strasburg, and in the Eastern States the "Silverskin," is the parent. These varieties are the Common Flat onion, the Early Cracker onion, the Danvers onion, and the Intermediate onion.

The COMMON FLAT (incorrectly called Silverskin in the Eastern States, a name which properly belongs to the White Portugal) is not so generally cultivated since the Early Danvers was introduced, as formerly. It grows to a diameter of about three inches, is compact in its structure, and of good flavor. It is a good keeper.

The EARLY CRACKER onion (see illustration) is very thin, of a beautiful honey color, quite compact, and oftentimes hollows a little at the bulb around the neck. It matures about a week or ten days earlier than the Early Danvers, and in fineness of structure and delicacy of flavor is unsurpassed.

The great practical objection to the cultivation of this onion on an extensive scale is the extreme care required in handling it; it needs care to prevent bruising and consequent rotting. For using in the fall, this objection would not lie against it with any great force; this and the Early Flat Red are excellent varieties for raising where the seasons are short. It grows to a diameter of from two and a half to four inches.

THE EARLY ROUND DANVERS YELLOW onion, was originated by Mr. Daniel Buxton and brother of South Danvers, by careful selections of the roundest and earliest specimens from the Yellow Flat onion. The Danvers is an early onion, maturing within about a week or ten days of the Early Red and Cracker onions. It is very prolific, and, like the Red Globe onion, gives larger crops by about one-third than the flat varieties. When each are seen just before pulling, the difference in the bulk of the crop is not very apparent, but when measured, the globular form of the Danvers "tells." When overgrown by too thin planting of seed this onion is at times rather coarse in structure, but ordinarily it is very compact, fine of structure, heavy, and a good keeper. When well ripened, I find it keeps equally well under the same circumstances as the common Flat onion. The earliness of the Danvers onion is a great gain in short seasons, or very wet ones; and as this onion begins to form its bulb quite early in its growth, ("bottoms down" is the farmer's phrase,) it presents marked advantages over the flat sorts for early marketing. In the Boston market the Danvers sells for somewhat more a barrel than the Red.

Having had considerable experience in selecting onions for seed purposes, I find that I can obtain a much greater proportion of handsome, well-developed seed onions from onions that have been raised from seed stock that has been carefully selected through a long series of years, and am led to believe that there can be "pedigree" onions as well as pedigree cattle, and that seed raised from them can be relied on under the same conditions to give a handsomer onion than can the average of seed.

WHITE PORTUGAL.

The cultivation of this early onion is mostly confined to the raising and planting of what are known as "Setts" or Button onions, or onions for early family use, as it is a poor keeper. It is a sweet, mild onion, of a good size for family use, though averaging considerably smaller than the varieties that have been described. Here let me say, that, for family use, except for frying, the common onions of the market are much too large to be economical,—the two outer layers of an onion four inches and upwards in diameter, though making up about half the bulk of the onion, are usually coarse and tough, and slough off when boiled. The sweetest, tenderest, and most economical onions for this purpose of the yellow sort are those that are from two to three inches in diameter.

THE QUEEN.

Of the newer sorts, the "Queen," a white English variety growing to from two to two and one-half inches in diameter, is doubtless the earliest at present known ; so early that under favorable circumstances it may grow to a market size about as soon as those raised from Philadelphia setts.

SOUTHPORT WHITE GLOBE.

This is a large, globular, white variety, that is about as round and, when cured in the shade, about as white as a snow-ball, being the handsomest of all onions. It is too late to be planted with safety in latitude north of southern Connecticut.

MARZAJOLE, MAMMOTH TRIPOLI, NASBEY'S MAMMOTH AND GIANT ROCCA.

Are European varieties that grow to a mammoth size in Southern Europe where for their mild flavor they are held in high esteem. Grown in this country they are of a milder flavor than our common sorts, but, though they grow larger, do not attain to the size they acquire in Europe, and though excellent for use in a green state are not good as keepers.

There are usually the distinctions I have here stated between the late and early varieties ; but some times drought and other causes will almost destroy these distinctions, ripening the very early and medium early sorts at the same time.

WHAT ONIONS SHALL I RAISE?

Having described the standard varieties, a beginner may query in his mind as to what variety would be most profitable and most reliable for him to cultivate.

The Danvers onion is the handsomest shaped, yields as much as any other sort, and more than any of the flat varieties, per acre. In the town of Marblehead, over nine hundred bushels have been raised on one acre of land. It is an onion very popular in the Eastern market and in Eastern Massachusetts is raised to almost the exclusion of any other variety. The Large Red onion is quite a favorite in the west, and is considered by some dealers to be the best variety for shipping purposes, though the Danvers is also shipped largely. Those who live in the latitude where the onion is difficult to mature from the seed in one year, affirm that the Red onion will mature farther south than any other variety.

After all, whatever suggestions may be offered, the local demand will do most for settling this point. Aside from this, I would recommend the Early Red Globe Danvers as, *on the whole*, the most desirable sort. The Red Globe is somewhat hardier than the Danvers.

ONION SETTS OR BUTTON ONION.

In that portion of the United States south of the vicinity of New York City, onions from seed raised as far south as the Middle States cannot be relied on to mature the first year, owing to the extreme heat of the climate forcing the formation of the bulb and drying down the top quite early in the season. *But if the seed was grown in the Northern States from carefully selected stock*, it will mature onions the first season when planted in the Southern States, as I learn from several of my correspondents, some of whom have grown them of market size the first season as far south as Texas. As a rule such onions are hardly as large as those grown farther north, but yet amply large enough for market. To give the rule concisely, if gardeners in the south wish to raise onions from the black seed so that they will grow to market size the first season, they should procure *seed grown as far north as possible;* and *vice versa,* gardeners in the North who wish to grow their own setts *should procure southern grown seed.* If in these southern latitudes two years are given to the maturing of the crop, the first year, the ground is prepared as already directed, excepting that it is but lightly manured; broad, shallow drills, from one to two inches in width, are made about ten inches apart, and these are sown early in spring, very thickly, at the rate of about thirty pounds to the acre, and the crop becomes mature in July, when it is pulled and stored in cool, airy lofts, being spread very thinly over the floor,—those raised from the White Portugal onion to a depth of about two inches, and those from the yellow sorts to a depth of about four inches. A gentle raking occasionally is of advantage to promote dryness and to prevent sprouting. The yellow variety is the best for keeping, and hence will bear the confinement incident to transportation with less injury; but the clean, white appearance of the onion raised from the white setts gives them the preference in the market. Attempts are often made in the north by market-gardeners to raise their own setts and thus save the large outlay often required to purchase them,—for most of the early onions now used in the northern cities are raised from the setts. The attempts to raise them in the north were formerly for the most part a failure; a large proportion of the setts so raised pushing seed shoots and thus spoiling the onion for market purposes, for the reason that northern grown seed was used. The true sett is an onion that has been checked in its annual growth and dried down before it has matured,—hence it has an additional growth to make before its annual growth is matured, and before this there can be no seed shoot pushed, for the onion is a biennial plant and the seed shoot belongs to the second year of its growth.

Setts are planted in rows about ten inches apart, and two or three inches distant in the row. As the ground worms are very apt to remove them when first planted, the bed should be occasionally examined. Some roll them immediately after planting, others hold to dropping them in shallow drills, not covering them at all with earth.

Onion setts vary in size from a pea to a hazel-nut. The smaller the size of the setts, the greater the number of onions contained in a given quantity; but many find it for their interest to purchase setts of a good size, as they yield larger onions. Among the market-gardeners in the vicinity of the large cities onion setts are very extensively planted, some planting as high as one hundred and fifty bushels annually. The quantity planted per acre varies with the size, from six to ten bushels.

RARERIPES.

Rareripes are onions raised by planting out bulbs of the growth of the previous season. The Rareripe oftentimes differs from the onion sett only in being a matured onion, as frequently they are about as small as the setts. The method of raising them is the same as that of raising early onions from setts, with the difference of planting them at times at greater distance apart in the row proportionate with their greater size. The raising of Rareripes is a very profitable way of disposing of such onions as are badly sprouted, are very small, or in any way unprofitable for marketing. A seed shoot may be uniformly expected from each onion; but as this greatly deteriorates the quality of the Rareripe, making it tough and woody in structure, it should always be cut off. If cut off before the swelled growth appears, (a striking characteristic of the onion family and a proof of the skill of the Divine Architect, in strengthening by so simple a process the tall, thin stalk designed to support the heavy seed head,) it will again shoot up; wait, therefore, until this swelling begins to show itself, and then cut below it, and no more trouble from this source will ensue. The smaller the onions planted as Rareripes, the handsomer will be the crop,—the very small ones producing each one handsome round onion, while the large ones produce two or more which are irregular in form.

POTATO ONIONS, TOP ONIONS AND SHALLOTS.

Potato onions, (see engraving,) Top onions and Shallots are thought by some to have originated from the common onion. It is certain that at times all three of these varieties are sported by the common onion. In a large field of seed onions, occasionally small onions will be found, growing in place of seed, and these onions when set out the ensuing spring will vegetate and develop readily, but they will not always in turn yield the like, *i. e.*, Top onions.

Potato onions, or multiplying onions, as they are sometimes called, are a thick, hard-fleshed variety, very mild and pleasant to the taste, and tender if eaten soon after gathering, but they grow to be tough as the season advances. They are poor keepers, unless spread very thinly in some dry apartment. They are propagated by planting the bulbs in drills, fourteen inches apart, the largest ones six, the smaller four inches apart in the row, and the smallest ones two inches. The small ones rapidly increase and make onions from two to three inches in diameter, while the larger ones divide and make from four to a dozen or even sixteen (usually from five to eight) small, irregularly shaped onions. It will be seen that the larger bulbs answer the same purpose as the seed in the common onion; hence to have onions for sale and yet maintain the stock, it is necessary that both sizes should be planted.

The Potato onion should be indulged for its best development in a soil rather moister than the varieties from seed. The advantage of the Potato onion is its earliness, and the fact that it is not as liable to injury from the onion maggot, when that abounds, as the common sort. I have seen an instance where, on half an acre of each growing side by side, the common onion (that raised from seed) was almost wholly destroyed, while the Potato onion was nearly uninjured.

Shallots differ from Potato onions principally in their characteristics of *always* multiplying; a Shallot *never grows into a large round* onion, but always multiplies itself, forming bulbs that average more oblong and are usually smaller than those of the Potato onion. I find them occasionally pushing a seed shoot, which I have never seen in the Potato onion. Their habit of growth is finer, making a longer and more slender leaf than the Potato onion. They are mild of flavor, and greatly excel every other variety of the onion family in their keeping properties: with little care they may be kept the year round. All seedsmen do not know the difference between the Potato onion and the Shallot. Within a few years I have twice had Shallots sent me under the name "Potato onion."

Top onions are propagated from little bulbs, which grow in this variety where the seeds grow in the common sorts. They grow to a large size, are pleasant, mild flavored, rather coarsely and loosely made up, and have the reputation of being poor keepers. Raised like the Potato onion.

SEED SOWING MACHINES.

There are a variety of machines in the market for sowing onion and other seed, but most or all of them can be arranged in four classes, *viz;* Brush Sowers, Snap Sowers, Drop Sowers and Agitators.

Brush machines are those in which the seed is forced out by a brush contained in the seed box. The characteristic feature in this class of seed planters is of English origin, and has passed through various modifications in this country.

The Brush machine, an engraving of which is here presented, makes the drills, drops the seed, covers and rolls it; it is adapted for planting all the common root crops.

The rows in this machine are marked out by a chain, two of which hang near the handles and drag on the ground, being used alternately. The wheel is pushed along the mark made by the chain.

The principle on which the "Snap" machine is founded is the securing the flow of seed through the aperture by a jerking motion, which is usually effected by a spring which makes a snapping noise when set free.

One of the machines built on this principle is known as the Danvers Onion Sower. (See engraving.)

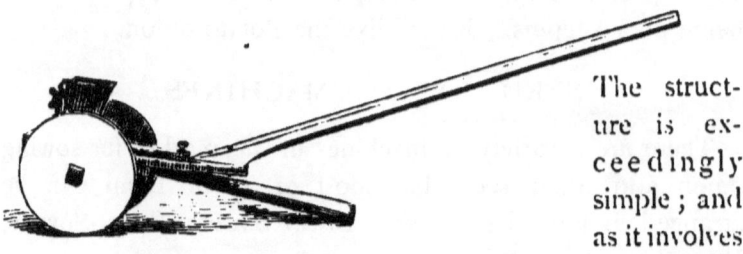

The structure is exceedingly simple; and as it involves but few parts, and hence is but little liable to get out of order, and when out of order can readily be repaired by any blacksmith, it was quite a favorite with onion-raisers, but improved implements have to a large degree taken the place of it.

The Danvers machine opens the furrows, drops the seed, covers it, but does not roll it. Farmers usually attach an old horseshoe to the end of the seed coverer, which gives sufficient weight to make it answer the purposes of a roller.

Matthews' Garden Seed Drill is a fine example of the seed planters on the "agitator" principle. This agitator is a finger of iron in the seed box which projects just over the orifice from which the seed drops and by a motion given it by the revolving of the wheel, keeps the seed continually stirred and thus prevents its clogging,—a trouble to every gardener when planting such seed as beet and parsnip. Were I called upon to recommend a seed sower for general work it would be the Matthews; all in the market have more or less of good qualities but I have found that the Matthews combines more than any other one.

On light soil hand cultivators are useful. These are now

 sold combined with seed sowers, so that the same implement may be used for either purpose.

The Matthews' hand Cultivator is a good illustration of this class. (See engraving.)

In these machines the seed falls through holes in little slides of tin, different slides being substituted as the seed to be sown is larger or smaller, or the quantity to be planted is greater or less. Farmers will often find it for their interest to enlarge or diminish the size of these holes. The holes in the tin of the Danvers sower, to give a liberal sowing of about four and a half pounds to the acre, should be large enough to drop ten to fourteen onion seed to each snap. By putting the hand under and counting the seed which falls in a dozen snappings of the machine, a reliable average can be ascertained. As the size of onion seed often varies, no particular size of hole can be relied upon; it must be tested for each season. Another convenient test is to trundle the machine over the barn floor, or a newspaper spread and secured in the field, and observe how thickly the seed fall. For a beginner the first test is the better one.

Of the sower which drops the seed in hills I will treat presently under the head of "Onions with Carrots."

PLANTING THE SEED.

Having selected our seed sower and regulated it, the next step is to plant the seed. It is exceedingly necessary that the first row planted should be straight, as this becomes a measure of straightness for all the others. A steady hand and a straight eye are of great value here; but with a little practice a good degree of accuracy can be obtained by most persons, though a few will always find it for their profit to hire some experienced hand. Two or three sticks may serve to mark out the first row, and by keeping these bearing on each other as the machine is pushed along, the first line must be a straight line. In some machines the chains which drag from the handle, and in others the wheels, serve to mark out the rows.

As the Scuffle Hoes (see engraving) and Wheel Hoes (see engraving) to be used will be of a constant width it is important that the width of the rows should be kept constant, particularly that they should not be brought nearer together than the distance fixed upon.

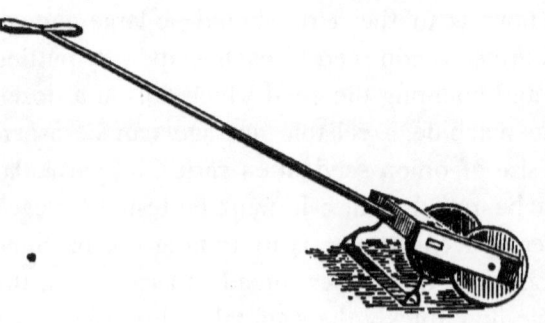

The distance between the rows varies in different sections from twelve to fourteen inches; when seed are planted for setts, ten inches is the usual distance between the drills.

The various hoes used in weeding are pushed before the operator and again drawn quickly back, the operator taking short steps, and making the hoe cut in both the forward and backward slides. After the tops get so far grown as to bend over into the rows, they are apt to be caught between the wheel and axle of the wheel hoe; hence from thence forward the scuffle hoe should be used.

Farmers sometimes make their own scuffle hoes out of a  piece of an old saw, the teeth answering a good purpose in cutting off the weeds. The V-shaped hoe, (see engraving,) called Howard's Patent, will do excellent service if a weight of about two pounds be fastened around the handle near the ground.

In Eastern Massachusetts fourteen inches is the usual distance between the rows; while in Southern New England and parts of the West, twelve inches is preferred.

Before planting the seed it should be carefully examined, to see that it is perfectly clean from small stones, or any substance that can possibly clog the hole of exit. Let it be remembered, when regulating the machine, that the seed will not be likely to fall so fast from a full hopper as they will when it is nearly empty. The seed should be sown from half an inch to large an inch under the surface. The lighter the soil, the deeper the seed may be sown. It is thought that deep sowing has the advantage of getting the plants so deeply rooted that they will bear having the earth slightly pulled away from them in the first weeding, without so much injury as sometimes results when they are planted shallow. While planting, as well as when using the hoe, our farmers will find the advantage of having a finely pulverized surface to work on, free of all clods, sticks and stones, as such will continually vary the straightness of the rows, interfere with the planting and covering of the seed, and, when the hoe is used, glance it out of its course in among the tender plants.

HOEING AND WEEDING.

In from two to three weeks, if the weather is an average for the season, the young plants by a close examination may be seen pushing their green arches above the surface, bearing a close resemblance to a curve of grass. As soon as sufficiently up to enable a sharp eye to determine the course of the rows, without delaying a day or an hour, if the weather permits, the prudent cultivator will slide through his Scuffle Hoe, as at this season of the year the weather is very uncertain, and the land may become too wet to be worked soon after the young plants appear, and yet not too wet to hinder a rapid growth of weeds. Sowing a few radish with the onion seed is sometimes practised. As the radish seed vegetates in a few days the rows are thereby marked out and the wheel hoe can be used earlier. Care need be taken that the radish seed are not larger than the onion and so clog the hole. To obtain very choice cabbage plants, which grow fine and stocky, farmers drop a few seed into the hopper with the onion seed. On so rich a seed bed, prompt action is very necessary, or a miserably discouraging tangle will soon be the result of negligence. In their comparative freedom from weeds the cultivators in the West, on their new land, have a great advantage over their brethren in the East. By selecting pasture land and avoiding the use of barn manure, the work of weeding may be greatly reduced. I have raised a crop on such land, when the entire expense from after the crop was planted until it was gathered and got into the barn was but $35.00 to the acre. It was so free of weeds that one man slid through, hand weeded and partly thinned an acre and a quarter inside of a day. In about a week after the hoe has passed through them, the young plants will need their first weeding with the fingers. This is hand-and-knee work and, pursued as it has to be in this position at intervals throughout the heat of summer, it is to many the most wearying work of the farm. Boys being more nimble fingered than men, besides working for lower wages, there is a great saving in employing them, *provided* they can be relied on to pull up the roots of the weeds. When several are at work it will be wise to have a man with them.

To protect the knees from sharp stones, "pads" are used, which consist of squares of about eight inches, of several thicknesses of woolen usually covered with leather, strapped to the knees. In ordinary seasons onions require three or four hand-and-knee weedings, and from four to six slidings with the hoe. A man's judgment must be his guide. As onions shade the ground but slightly, weeds grow rapidly in onion beds ; and if they are once allowed to get the start, the labor of cultivation is immensely increased. Some cultivators practice scratching the soil away from the onions when weeding, with an old knife curved at a right angle near the point, or by a piece of iron hoop curved, the end being nailed to a small piece of wood conveniently held in the hand. Others practice throwing the soil slightly around the young onions with a scuffle hoe made with reference to this use, with a view of smothering the small weeds. Noyes' hand weeder will be found a very handy little implement for removing weeds, particularly

when the surface of the ground is somewhat hard. When the onions have begun to "bottom down," *i. e.* form their bulbs, it is the general practice to remove as far as practicable any surplus earth that has accumulated around them. In weeding, two or three rows are taken together, the weeds being dropped between the rows. Just before the crop ripens down, larger weeds will show themselves here and there over the beds; these are generally gathered in baskets and dropped at the end of the rows. If the seed of such weeds get ripe before they are pulled, the weeds should be carefully deposited in a pile in some by-place, where they can be burned when dry. Growers who practice throwing such weeds to their hogs because they are large and succulent, make an annual seeding of their beds with weeds. Particularly is this true of Purslane, one of the greatest plagues in the heat of the season. The habit of this plant is to ripen the seed, well down on the stock, while the main body of the plant is in its full vigor; hence it usually happens that much seed drops into the land some time before it is pulled, while the farmer never mistrusts it has ripened. I have seen Purslane completely eradicated from garden plots where it formerly was a pest, by a little care in this matter of letting it go to seed. The same remarks apply to the weed known as chickweed. When blank spots occur from poor seed, poor planting, or the ravages of the onion maggot, bush beans, cabbages or tomatoes may be planted.

When the plants are too thick, over one to two inches, they should be thinned; but the beginner had better pull with a sparing hand, for, if the ground has been manured very liberally, the crop will do well when the plants are very thickly together, and they will oftentimes grow as large when very thick as they will with three times the room. Onion-growers like to see their onions piled two or three deep as they grow, the upper layer being entirely out of the ground with the exception of the roots. When the tops begin to fall over, the onion is rapidly maturing, and the bulbs will now grow very fast. Farmers will tell you that "the top is going down into the bottom." The Flat onions begin to bottom late in the season, while the Danvers makes a very encouraging show of bulb quite early. Should the land have been but poorly manured in seasons of drought, the crop will be apt to be ripened prematurely, forming a small sized onion, while (divided it may be by merely a wall) those that have been more liberally manured stand the drought, and keep green sufficiently long to receive advantage from the later rains; an investment of twenty dollars in manure thus making a difference sometimes of a hundred dollars in the crop. If the crop is quite backward, late in the season the necks of the onions are sometimes bent over to hasten the formation of the bulb. This is done by hand, or by rolling a barrel over two rows at a time.

STORING THE CROP.

When the necks have fallen over, and the great proportion of them are dry, the crop should be pulled by hand and be laid in winrows, about three rows being put in one. At this time all weeds remaining should be pulled and piled, preparatory to the final clearing of the bed. The pulling of the crop should not be delayed after the tops are well dry; for if rain should now fall, the onions will be apt to re-root to their injury. Should the backwardness of the season make it necessary to pull the crop in rather a green state, it will be well to allow it to remain untouched, after pulling, for about a week, before turning or stirring, which will tend to hasten the decay of the greener tops; otherwise they should be carefully stirred every pleasant day with a wooden-toothed rake. This should be very carefully done, as the onions are very easily injured, especially at this stage, and when injured are almost as likely to rot as a bruised apple. See that they are not injured by the raking or treading of a careless hand. When the crop is thoroughly dried, the onions feeling hard to the handling, it will be ready for topping for market. They are carefully collected in baskets, rejecting all stones, scallions and rotten onions, and taken in wagon loads to the barn, when the tops are cut off clean to the onion with a sharp knife, or if the necks are small and quite dry they may be pulled off by the thumb and finger. This is usually done by boys or females from two to four cents a bushel. While collecting, look sharply on the bottom of the onions to detect rotten ones. Some growers prefer to leave such of the crop as they design to keep for a late market untopped. If it is intended to market the crop immediately, the onions may be piled to a depth of three or four feet; otherwise they should not be over two feet in depth. Leave the barn doors and windows all open every pleasant day. As the crop is topped, those of the size of a hazel-nut and smaller are classed as pickle onions, these being marketed principally for that purpose, and usually bringing about half the price of the full grown ones.

TRACING, OR ROPING ONIONS.

When the crop has ripened down but poorly, the greener onions are at times traced. This is done by cutting off the neck within about two inches of the bulb, and binding it to a handful of straw; beginning at the butt end of the straw, lay the neck against the straw, give two or three firm turns with the twine (net or wrapping twine), add another onion, and thus proceed till the straw is covered, the larger onions being tied to the bottom and gradually decreasing in size to the top Onions so slowly ripened that they would soon spoil if stored in a mass, will keep well when traced, and oftentimes bring a greater profit than the best of the crop. Rareripes, and such of the earlier onions as are to be sent long distances, or be kept awhile before marketing, are sometimes traced. Traced onions keep in good condition a long while in a dry, cool place. Within a few years tracing has almost ceased in this vicinity.

MARKETING THE CROP.

The Sett onions, Potato onions, Top onions and Rareripes, in some sections, are for the most part sent to market in a green state in bunches. The Potato onions are brought from the South, dry, in large quantities to supply the Northern markets, soon after the arrival of the Bermuda onions, just before the ripening of the Northern crop. After the Potato onions follows the earliest variety of the Red, and immediately after, the Danvers, and finally, the large Red completes the season. The sales in the Northern markets early in the season are made mostly for the supply of the local immediate demand, the great bulk of the crop not being sent in before the call for shipping purposes has commenced. For this reason, farmers find it to be for their interest to do but little more than feel the market until about the middle of October, as large purchases made previous to this period are mostly as an investment by speculators, with the exception of such lots as go to supply the markets of large towns and cities of the extreme North beyond the limits of the onion-growing region.

The price of onions varies greatly; they have sold as low as seventy-five cents a barrel, while the early crop of 1864 sold as high as sixteen dollars a barrel, by the five hundred barrels. From September to March, in the same season, the fluctuation is sometimes between two dollars and six dollars. Crops have at times been sold to be delivered in the course of two months, and in that time have more than doubled in price. The general truth is, that those brought latest to market, being kept till near spring, bring the best prices. The great facilities afforded for onion raising by the fertile soil and favoring climate of the West, will doubtless in a few years tell powerfully on the Eastern market.

PRESERVING THE CROP.

If it is the design to keep the crop for a winter market, it should be stored in a *cool*, *dry* place, out of danger from severe frosts, in bulk, but not over two feet in depth. Onions will bear a temperature of 28 degrees without injury, or any degree of cold if kept frozen till the final thawing when in bulk. It is a good plan to let them rest in a lattice work of slats on the sides and bottom of the building, that the air may circulate through them. If kept in barrels, these should not be headed, and should have two or three openings made with a hatchet or large auger in the sides near the bottom. If it is designed to keep the onions till spring, the cheapest and best way is to freeze them. To do this, select the northwest portion of some out building under which the air does not circulate, spread the onions about one and a half feet in depth, leaving a vacant space of about two feet from the side of the building, let them get thoroughly frozen, then cover them closely with an old sail, or any cloth, to keep the hay from mixing with them, and spread the hay two feet or more in depth above the covering; also pack fine hay closely between the heap and the sides of the building. Here let them remain untouched until the frost is entirely out, when they should be spread at once, well aired, and turned carefully and often until thoroughly dried.

If the onions in the fall are not well ripened, or if a larger proportion than usual are rotten, which is apt to be the case after a very wet season or when the onion maggot has given much trouble, to store largely for winter sales is attended with great risk. I have known one enterprising cultivator to sink several thousand dollars in a single season by storing heavily under such circumstances.

SMUT, RUST AND MAGGOT.

The onion crop is sometimes severely injured by a disease resembling mildew. The tops of the leaves die, and the whole plant is more or less covered by this white blast. From the effects of it the onions almost cease their growth, and the crop finally obtained is small in size. This disease in some sections is known by the name of "rust." It is more frequent in extraordinary wet seasons, and is more common on old beds than new. The best remedy yet known for old beds is to carefully remove from the bed and destroy at the close of the season all diseased bulbs, as they will be likely to spread the disease by giving it a lodgment in the soil, then run the plow a little deeper, and thus mix in a little new soil.

The onion maggot is hatched from the eggs of a fly, which are deposited in the plant (not the seed) very near the surface of the ground. Its presence may be detected in the crop when very young by the sudden turning yellow and falling over of the plant, when, if the attempt is made to pull, it will usually break off near the surface, and on squeezing several small maggots will present themselves. Some writers state that the fly deposits its eggs only at an early period in the growth of the plant. It is true that some seasons the injury is most marked previous to the bottoming of the onion, but I have seen beds injured at every stage of their growth, and in one season about half of the crop was destroyed by the maggot at the close of the season after the onions had been pulled. Various remedies have been proposed, but of these it may be said that they are not practical on a large scale. The idea on which most of these is based is that of producing a scent so disagreeable as to drive away the fly; but old experimenters recall the capacity of the Canker-worm moth and the Squash beetle to ignore the most repulsive obstructions of this kind when stimulated by their instinct to deposit their eggs. Pine sawdust, either clear, soaked in the urine of cattle, or in the ammoniacal liquor from gasworks, scattered over the bed just before the appearance of the plants, at the rate of a bushel to ten square rods, guano sprinkled along the rows and on the plants, twice during the season, unleached ashes used in the same manner,—these have given satisfactory results to some growers. Scalding water poured from a common watering-pot through a hole the size of a pipe stem, along the drills near the roots of the plants, and repeated three or four times during a season, is said to be efficacious. It is obvious that the practical value of such a remedy must be confined to a very small area of land.

In New England the maggot has been slowly making his way North, adding greatly to the uncertainty of the crop, until his ravages have extended to Southern Massachusetts. Very light soils appear to be most affected by his ravages. In some seasons the injury done is insignificant, and on the whole the area planted in Massachusetts has not been materially reduced.

He will one year confine his ravages mostly to one portion of a township, and the next season reverse matters; while some tracts are almost never injured, on others he appears to settle down as a permanent resident.

RAISING ONION SEED.

What does all this investment of money, time, labor and watchfulness, amount to if the seed is worthless, has no vitality, is not true to name, or was grown from worthless trash? Onion seed should be raised from the very best onions of the very best crop grown in the vicinity. The best type should be first selected, which should be a medium sized onion, very hard and compact in structure, with a close, thin, fine skin, and a very small neck. Those selected for seed should be the earliest ripened of the crop, provided such are fully ripened and not blighted. To select the earliest onions, the seed grower should visit the field before the crop is pulled.

Onion seed is sometimes (I fear too often) grown from the entire crop, be it good, bad or indifferent. A great step of improvement on this is to purchase outright as good a crop as can be found; but the only way to secure and keep the best and most reliable seed is that first given. Poor onion seed is always very dear indeed, as a present, while first-class seed at the highest price yet paid is worth a long and careful seeking.

Seed onions should be kept in a cool, dry place, spread to about a foot in depth; if kept in barrel, (old lime casks are best) these should be left unheaded, and two or three pieces should be chopped off near the bottom to admit a circulation of air. As early in the spring as the ground can be worked, they should be set out in trenches, (the onion when covered in trenches will stand a heavy frost without injury,) which should be from three to four feet apart and about four inches below the surface, the land having first been heavily manured. Some good seed growers apply their manure directly in the trench, while others spread it broadcast and plow in. I prefer to plow in a liberal quantity, and then use ashes, superphosphate of lime or guano, in the rows, applying it just before covering onions. If the onions are much sprouted, the sprout may be cut off quite home to the onion, which will insure a straighter and healthier growth. Care should be taken to plant right end up, for, odd as it sounds, in the spring it sometimes requires a little care to determine which is the right end. As soon as the onion is well rooted, the earth should be drawn up to it; and this should be done three times during the season, until the earth is heaped around them eight or ten inches above the surface of the ground. The first hoeing should be given them very soon after the sprout starts, to *fully cover the onion*, as when exposed it is very apt to decay. With this support, on land that is not too moist, I find that no further precaution is necessary to keep the seed tops from the ground, though it is the practice of many growers to support with light strips of wood, or a line drawn along about two feet from the ground. After the last hoeing, (and very clean culture should be given then), they should be gone among as little as possible.

The seed tops may be safely cut (leaving about six inches of the stem on) when the seed vessels begin to crack; or what is a better guide yet (for after the seed vessels begin to crack much seed is apt to be lost, especially by heavy storms,) after the turning yellow, near the ground, of the seed stalk; when this occurs, the top may be removed immediately, even though it should appear quite green above.

Seed tops will be often found in which the seed in the shortest-stemmed receptacles is ripened, and the receptacles themselves are cracked, while a fresh growth of seed vessels in a green state almost conceal them; in such cases I would advise the cutting of the top. The tops when cut should be spread to a depth of six inches or less, in a warm place where heat and air abound, and be turned two or three times daily, until thoroughly dried, when the seed is ready to be threshed out; or it may be stored in barrels in a dry loft, and threshed as wanted. If the seed is plump and has been well ripened, the frequent turning of the stalks will have shaken out by far the larger proportion of it,—in some seasons more than five-sixths.

As the seed stalks make but little shade, the ground between the rows can be cultivated to spinach, lettuce, radishes, turnips, or some early vegetables, then this will make the hilling of the seed more costly, and when these are harvested, be planted to cucumbers for pickles. The planting between the rows should be confined to the middle, and in trenches an inch or so below the surface, unless it be made after the onions have received their final hoeing; otherwise the drawing of the earth around the seed stalks will seriously interfere with these crops.

Strange as it may seem to those who have not tried it, such rampant growers such as squashes can be raised among seed onions and generally with no material injury to the seed. I have known five tons of Hubbard Squashes grown on about half an acre of ground planted to seed onions. The squash should be planted towards the close of May, after the onions have received their final hilling, two or three seeds being planted close to every other row, and about nine feet apart in the row; allow but *one plant to grow in a hill*. The vines, thus having plenty of room between the rows to spread about, do not incline much to climbing on the seed stock. Care should be exercised to break off at once the tendrils of such as attempt to climb. The one plant to a hill system will be found to yield as liberal a crop and finer squashes than the old system of three or four to the hill.

The yield of onion seed to the barrel of seed onions varies greatly; indeed, no investments near the seaboard prove more speculative. The maggot sometimes proves very destructive, so much so that the crop will not average half a pound to the barrel, while under very favoring circumstances crops have been raised which average eighteen pounds to the barrel.

When the seed is fully dry, (and seed that has been sunk should have a long exposure to the air and frequent stirrings; I have known large lots spoilt from want of care in this), it should be so stored as to be safe from all injury from cats and other animals, who are apt to resort to it, to the utter destruction of its vitality.

RAISING CARROTS WITH ONIONS.

The plan of raising carrots with onions is considered a great improvement by many who have adopted it, as the yield of carrots is thought to be a clear gain, diminishing but little or none the yield of onions. Carrots are planted in two ways; one by sowing them in drills between every other row of onions, and the other, which is considered an improvement, called the Long Island plan, by planting the onions in hills from seven to eight inches from center to center, dropping a number of seed in each hill, and from the first to the twelfth of June planting the carrot seed, usually by hand, between these hills in two rows, then skipping one, and thus on through the piece. The onions as they are pulled are thrown into every third row, the carrots being left to mature. By this method from two hundred to six hundred bushels of carrots are raised per acre in addition to the usual crop of onions. More manure is required for the two crops than for the onions alone.

The Machine used for sowing in drills has two boxes attached to the axle at equidistance from the wheels; there are three or four holes in the axle that communicate with the seed in the boxes, and as these holes pass under the boxes they are filled with seed, and as they turn the seed are dropped into the earth. Screws are sunk into the holes, which can be sunk more or less at pleasure, and the quantity of seed which the holes will contain is thus graded.

The machine should first be tested and so regulated that on a barn floor it will drop from eleven to twelve seed from each hole. When so regulated, on using in the field it will drop but from seven to twelve, owing to the more uneven motion.

This, like all sowing machines, and the same may be said of the scuffle hoe and wheel hoe, is pushed along before the operator.

My farmer-friends, I have now given you the result of my own experience in the raising of onions, Potato onions, Top onions, Shallots, and onion Setts, combined with the experience in onion growing of a neighborhood where a hundred thousand bushels are raised annually, with the results of personal observation in other localities, and with facts that I have collected by corresponding with different sections of the United States.

I hope this contribution will prove acceptable.

JAMES J. H. GREGORY.

MARBLEHEAD, MASS.

ANALYSIS OF THE ONIONS.

A recent analysis, under the direction of Prof. Goessmar of the Massachusetts Agricultural College, gives the following as the principal constituents of the onion:

Air dry onions without leaves were found to consist of:

Water (at 100° to 110° C.)	89.20 per cent.
Dry vegetable matter	10.80 "

and contained the following amounts of—

Nitrogen	0.212 per cent.
Sulphur	0.048 "
Ash	0.436 "

The percentages of the principal constituents of the ash were:

Potassium oxide	38.51 per cent.
Sodium oxide	1.90 "
Calcium oxide	8.20 "
Magnesium oxide	3.65 "
Sesquioxide of iron	0.58 "
Silicic acid	3.33 "
Phosphoric acid	15.80 "

Sulphuric acid not determined.

In the above table Potassium oxide, Sodium oxide, Calcium oxide and Magnesium oxide, mean practically pure Potash, Soda, Lime and Magnesia.

I infer from the table that of the three grand essentials in manure, Nitrogen, Potash and Phosphoric acid, the onion requires about equal quantities of the two former and half as much of the latter.

I trust this contribution will prove acceptable.

JAMES J. H. GREGORY.

MARBLEHEAD, MASS.

SQUASHES.

HOW TO GROW THEM.

A PRACTICAL TREATISE ON SQUASH CULTURE, GIVING FULL DETAILS ON EVERY POINT, INCLUDING KEEPING AND MARKETING THE CROP.

BY

JAMES J. H. GREGORY,

MARBLEHEAD, MASS.

NEW-YORK:
ORANGE JUDD & COMPANY,
41 PARK ROW.

Entered according to Act of Congress, in the year 1867, by

ORANGE JUDD & CO.

At the Clerk's Office of the District Court of the United States for the Southern District of New-York.

LOVEJOY & SON,
ELECTROTYPERS AND STEREOTYPERS.
15 Vandewater street N. Y.

INTRODUCTION.

The recent great increase in interest in squash cultivation, which has been promoted by the introduction of new varieties, has seemed to me to demand a more thorough and exhaustive treatment of the subject than is to be found in our present standard works on horticulture or agriculture. I am sustained in this position by the great number of questions propounded to me annually in the course of an extensive correspondence. To answer these questions, and to bring so delicious a vegetable as the squash into a more general and more successful cultivation, is the object of this treatise. The Squash family (*Cucurbitaceæ*) have their habitat in the tropics and warmer portions of the temperate zones; hence they require our hottest seasons to develop them in perfection. With the exception of the Vegetable Marrow, the squash family is almost unknown to our English cousins, as likewise is true of our corn and beans, for though the average temperature of the year is higher with them than with us, yet the extreme hot weather, which these vegetables require, is there wanting.

The introduction of the squash is a matter of the past half century; until within that time, with the exception of the Crookneck, the pumpkins, yellow and black, or "nigger," were the only varieties cultivated. Though the appetite for squash appears to be in a considerable degree a matter of education, yet it is becoming more and more popular in the vicinity of the large cities of the North, where among vegetables, it now ranks next to the potato.

WHAT IS A SQUASH?

In many parts of the South and West, where the fall and winter squashes are not much cultivated, the term "Pumpkin" is used for all the running varieties of the squash or pumpkin family, with the exception of the "Cushaw" class, which includes varieties that are closely allied to the Crookneck. To clearly define what is meant by the word squash in contradistinction from the word pumpkin, as used among market-men, is no very easy matter, as all the varieties, with the exception of the Crooknecks, easily intercross with each other, and in the recently introduced Yokohama, I have reason to believe we have found the connecting link between the Crooknecks and other squashes, thus destroying the reputation which the Crooknecks had hitherto enjoyed of being *the* squashes of the squash family. Grouping all the running varieties together, we express the marketman's idea of a squash, as distinguished from a pumpkin, when we say that all varieties having soft or fleshy stems, either with or without a shell, and all varieties having a hard, woody stem, and without a shell, are squashes; while all having a hard stem and a shell the flesh of which is not bitter, are pumpkins; and all of this latter class, the flesh of which contains a bitter principle, are gourds. In a more general classification, all varieties having a hard shell, are gourds, and those without a shell, are squashes. I had an amusing instance under this system of classification in a lot of seed, ordered from France as "gourds;" on examining them, I found that several of the kinds were varieties of our table squashes. Making a separate classification of the summer varieties, I define such to be squashes, in contradistinction from gourds, as are eatable at any period of their growth. It will be seen that the distinctions I make are more commercial than strictly scientific. What I aim at, is, to so define squashes, pumpkins, and gourds, that experienced market-men, seedmen, and new beginners, may meet on common ground, and clearly understand each other when using these terms.

In passing, I remark, that gourds are far more prolific than either squashes or pumpkins; in some instances more than two score having been grown on a single vine.

SELECTING THE SOIL.

All of the family thrive best, other things equal, in a warm soil, which is a soil through which the roots can easily find their way. The Hubbard squash appears to attain to its highest development in regard to both yield and quality in a soil, that, in addition to being warm, is also a *strong* soil. I would not advise planting in a clay soil, unless it be possible by thorough draining and high manuring, (for this purpose, long manure is better than fermented,) to make such soil light and porous. A drained meadow will often yield enormous squashes, if well manured, but they are apt to be very porous in their structure, of poor quality, and poor keepers.

Some years since I planted a piece of rich, black meadow to Hubbards, after manuring liberally in the hills. The result was a tremendous growth of vine, some of the leaves measuring twenty inches in diameter, while the ends of the runners, in their great vigor, lifted themselves by thousands two and three feet above the surface, and with their blunt, arched extremities, looked like a myriad of huge-winged serpents running a race. The squashes were of a light green color, very large and showy, but, when gathered, proved light in the handling, very porous in structure, cutting like punk, were very poor keepers, and coarse and watery in quality. Though such meadows are thoroughly underdrained, the squashes grown on them are light in proportion to their size, (which always insures poor quality and poor keeping,) unless the meadows have had abundance of sand and loam worked into them, thus adding the proper proportion of silica to the vegetable humus. Some years ago, when the Marrow squash was a novelty, bringing about $4.00 a hundred pounds, one of my townsmen raised some acres on a piece of drained meadow. Only a portion of the meadow had received a good dressing of sand; here the squashes were of about the ordinary size, while on the remainder they grew "as big as barrels." He traded a part of the crop with a peddler for a lot of swine. When the peddler called for the squashes, agreeable to instructions, the father being absent from town, his son showed him the smaller sized lot, saying that he had received directions to deliver them, as they were the best of the crop. But the peddler declared that, as he had supplied good pigs, he was entitled to good squashes, and would be put off with no trash. He therefore loaded his wagons with the "big as a barrel" lot, and left for home. Before many days a friend called, and, with a laugh, asked if he had heard of the result of the squash investment. "There was'nt enough substance in them to hold together until he got home'; they were carried to market in a few days, and two tons out of five were rotten." If the soil be wet and cold, the growth of the vine is much retarded, and not only is the crop much lessened in size and weight, but at times this singular result is seen—the squash loses its normal form. I have seen a crop of Hubbards grown under such circumstances, all of which were nearly flat at each end, instead of having the elongations that belong to the normal form.

When two soils of equal natural strength, but one of them being more gravelly in its structure, are heavily and equally manured, I have noticed, in several instances, that the more gravelly piece will give more squashes, and less vine than the others.

Unlike some varieties of melons and cucumbers, squashes will do finely on freshly broken sod, which has the advantage (a great one in many localities) of being less infested with bugs, than old tillage soil. The practice of digging holes a foot or two in diameter in patches of turf in waste places, around hedges, or in corners of fields, which, after filling with manure, are planted to squashes, is but a waste of time; the result is, a growth of vine of a few feet in length, the setting of squashes, and then both squash and vine become checked in their growth, as the roots of the vine make vain efforts to penetrate a dense mass of hungry grass roots in search of food, the leaves gradually turn yellow, and before you know it, have entirely disappeared. By pulling on a dead vine, you drag out a half grown squash hidden among the grass.

If the sod abounds in the pest known by various names, as witch, twitch, or quack grass, there is some danger that the grass will overrun the vines. If the grass has not been quite thoroughly torn up by the cultivator before the vines begin to run, better plow up at once, as the crop will be nearly a failure. Hoeing up and hand pulling the grass will practically amount to nothing under such circumstances, as I once learned to my sorrow. If the sod is not very badly run to twitch, there is but little danger, provided the cultivator is faithfully used from the time the vines appear above ground until the runners begin to push.

THE MANURE.

The squash vine is a rank feeder. Night soil, barn manure, wood ashes, guano, muscle mud, hen manure, superphosphate of lime, pig manure, sheep manure, fish guano, fish waste—either of these alone, or in compost, is greedily devoured by this miscellaneous feeder. The great error in the cultivating of the squash is to starve it. By many cultivators, when every other crop has had its share, and the manure heap has been used up, a piece of sod is broken for the squash patch, about the only food depended on for the crop being what it can gather from the decay of the fresh turned sod. Under such treatment, the crop is small, the squashes small, and the general result unsatisfactory. Another error of the opposite extreme is one often committed by market gardeners, who have learned that no paying crop can be grown without liberal feeding —who give all the food necessary, but do not allow sufficient room for the extra growth of vines under such culture. Of this latter error I propose to treat under the head of "Planting the Seed."

Night soil should be used, mixed with muck and other manures, in the form of a compost. It may, however, be applied fresh, directly to the hill, if sufficient care is taken to mix it thoroughly with the soil. Some years ago, I broke up a piece of land in the spring of the year for squashes, and the location being difficult of access, I used night soil from a vault on the premises, pouring about two bushels into each hill. After we had finished manuring, I sent my hired man, stout Jim Lane, around with his hoe to mix it thoroughly with the soil in the hills. When Jim came back, saying the thing had been thoroughly done, I send him around a second time, to give it another mixing up, and, on his return, sent him around the third time, though the old fellow assured me that it couldn't be improved on, and I had no doubt he had done his work well each time, but, with two bushels of fresh night soil in each, I knew that all the danger lay in one direction. The result was, the vines came up a rich, dark-green, and took right hold of their food.

With the exception of barn manure, it is necessary that each of the manures mentioned above should be well mixed in the soil when used in the hill. When wood ashes are used, they should not be mixed with other manure, until just as it is applied, as this would injure the value of the manure, by setting free the ammonia. When I have used ashes in connection with Peruvian guano, I have been in the habit of putting layer with layer in a wheelbarrow, hurrying it to the hills, and then covering it immediately with soil. Even with all possible hurrying of matters, the strong, pungent smell of the escaping ammonia could be readily detected.

Wood ashes, mixed with fresh night soil in the hill, is considerably worse than nothing. Some years ago, aiming to grow some extra large specimens, I selected a favorable location, opened several large hills, and poured into each about a couple of bushels of night soil. Into this I stirred a liberal quantity of wood ashes, acting on the theory that its alkaline properties would serve as a corrective of the rank crudeness of the night soil. I pulled the earth over the hills, and planted my seed. The seed vegetated, but the young plants soon came to a stand still. I applied a little fresh soil to the roots, thinking the manure below might be too strong for the young rootlets to absorb. Still, there was no growth; soon the leaves turned yellow, and the plants died. I opened one hill to find the cause, and there I found cause enough in the presence of a mass having about the size and appearance of an ordinary grindstone; the ashes and night soil in combination had made a hard cement, and the entire contents of each hill could be rolled out in one cake.

HOW MUCH MANURE?

Those who, under the stimulus of a city market, follow market gardening, soon learn one truth that may be set down as an axiom for successful gardening, viz.: that other things equal, it is the last cord of manure that gives the profits. There is but very little danger of giving too much manure to your squash ground, provided the hills are made at a proper distance apart, and the vines are not too numerous.

No prudent man will plant squashes with less than four

1*

cords of barn manure, or its equivalent, to the acre; this is the *minimum*—when squashes are raised as a profitable crop, from six to twenty cords of good manure per acre are used.

Twenty cords to the acre will, I doubt not, sound like a large story to many readers, and it *is* a large quantity, even for the high culture required for successful market gardening, but I have seen that quantity applied, and once, in my own practice, applied thirty-five cords to a little over two acres of squash land, where the soil had been over-cropped, (or rather under-fed,) for many years before I came into possession of it. Let us look a moment into that axiom—"the profits come out of the last cord of manure." With four cords of good barn manure to the acre, on *good* soil, the average yield would be about four tons of Hubbard squashes; with six cords of manure, the average yield would be about six tons; with eight cords, the yield would be from seven to eight tons. These are real results, that I have had in my own experience. Here it will be seen that we gain about a ton of squashes with each extra cord of manure; in other words, by investing eight or ten dollars, we treble or quadruple our money in six month's time—quite a profitable bank of deposit is the manure heap! Not only is the crop heavier, but the squashes are larger, and, therefore, far more marketable and, usually, at a higher figure, often readily bringing $5 or $10 a ton advance in the market. Nor is this all; the virtues of the manure are not exhausted the first season; but the ground is left in higher condition for the crops of the next season. Again, let it be noted that the cost of cultivation of a poor crop is just as great as the cultivation of a large one, while the promise of a large crop is a great cheer amid the labor of caring for it. The strongest argument for the liberal manuring of this and all other crops is, that a certain portion of the crop but pays for the cost of producing it, and that the profits can only come *after* the cost of production is paid.

The cost of producing an acre of squashes, independent of the cost of the manure, will be:

Plowing	$ 6.00
Distributing Manure	5.00
Cultivating in Manure	3.00
Seed	4.00
Mixing Manure in Hills	2.00
Planting Seed	1.00
Three Cultivatings in course of season	5.25
Two Hoeings	3.00
Lime and Liming	1.50
Hand-weeding of large, scattered Weeds, after Runners have started off	1.00
Gathering of Crop into Heaps ready for Carting	2.00
Interest on Land	9.00
Wear and Tear, and Incidentals	2.00
Total, exclusive of Manure	$44.75
Add cost of four Cords of Manure, at $8.00, landed in Field	32.00
Cost of Guano, or some equivalent, to mix in Hills	5.00
Total cost of Crop when four Cords of Manure are used per Acre	$81.75

Now, as we stated above, the average yield of Hubbard squashes, under such manuring, would be about four tons. The average price of Hubbard squashes in the Boston markets, for the past four years, of such a size as four cords of manure to the acre would produce, has been about $25 per ton. At this rate, the returns (not deducting the cost of marketing) per acre would be $100, from which deducting the cost of production, $81.75, we have $18.25 as the profits on the acre.

If, now, by adding two cords more of manure, or $16.00, to the cost of production, we obtain two tons more squashes, then the income is increased $50, (this supposes that we get but the same price per ton, but, in fact, I get from $5 to $10 more per ton for such squashes,) and we have a profit of $52.25. The two cords of manure extra have nearly trebled the profits; in other words, by adding about *one-six* to the cost of production, we treble the profits. Or, again, to give a commercial look to the matter, for every dollar invested in manure in May, in October, or five months, we receive a return of three dollars and an eighth. The returns have proved in the same proportion up to eight cords, and at times up to ten cords, to the acre. These statements are not visionary; they are drawn directly from *practical experience*, and can be corroborated by any farmer who has tried liberal manuring. Catch a farmer of that class going backwards, and putting less and less manure on his grounds, what a phenomenon he would be! No; the progress of all enterprising farmers is in one direction. By extra manuring the probabilities of receiving paying returns, are far greater in agricultural than in commercial life, as figures will readily show, though the popular belief is directly the contrary.

PREPARING AND APPLYING THE MANURE.

As a general rule in farming, the value of manures that are good for any crop, is increased by mixing them together, making what is called a compost. Ashes and common lime are an exception to this rule; each of them sets free the ammonia, (the most valuable portion of any manure,) and, being volatile, it escapes into the atmosphere. In preparing a compost for squashes, the bottom of the heap may be made of muck that has been acted upon by the frost, sun, and rain of a year, if practicable; if this can not be done, let it at least be got out the fall previous, that it may be disintegrated, and, in a measure, sweetened by the winter's frost. In the course of the winter, manure from the barn-yard may be hauled upon it. If this has been well worked by hogs, the better. Toward spring, if night-soil can be poured into it, the richness of the heap will be much increased. Sharp sand can now be thrown over the heap, and about as soon as frost breaks ground, the entire mass should be thrown over with forks, and thoroughly commingled, all coarse lumps broken up, and all frozen lumps brought to the outside of the pile. As soon as the mass begins to heat, the process should be repeated once or twice, until it is made as fine and as thoroughly mixed together, as time will allow. The sand will be found to be excellent to keep the manure finely divided and light, or to "cut" it, as farmers say.

In applying the manure for this or other crops, many farmers use all the manure in the hill; some, because having but little to use, they wish to get it as near the plants as possible, while others seem to hold the theory, that a circle of three or four feet in diameter is a sufficient area for the roots of squash vines to travel over in search of food. Where all the manure is used in the hill, the squash vines push over the ground rapidly, until just after the setting of the squashes, when they lose vigor, the squashes develop but slowly, and in the end there is a small crop of undersized squashes, for the roots, having meanwhile pushed beyond the hills, can not find food sufficient to sustain the growth of the vines. The roots of squash vines increase faster than is generally supposed. There is a theory that the roots grow to the same length as the vines, keeping pace with them in their growth. Whether the roots grow as long, or longer, than the vines, I can not say, but when the runner of a vine had pushed out but eighteen inches, I found the root over three feet in length, thus proving that at one period of growth, the root increases faster than the vine. This spreading of the roots through the soil is one of the marvels of vegetable life. I remember once lifting a small pile of litter that was about six inches deep, some dozen feet distant from a squash hill, when I saw what appeared to be a fine mist at the surface of the ground, but upon examination myriads of fine rootlets were seen, that were doubtless feeding on the decaying vegetable matter. Any person who will examine a squash vine of the running sorts, after it has set its fruit, will find roots pushed down into the earth at each joint; and though these may be in part designed by the Creator to steady the vine, there can be but little doubt but that they are designed also to feed the long runners. And this is proved by the fact, that if the connection of the vine with the main root be severed, while these subordinate roots remain uninjured, it will still maintain a degree of vigor. Such facts as these sweep all theories of hill-manuring by the board, for if the roots travel beyond the hill in search of food, there a wise cultivator will put food for them. My usual practice is this: to distribute all the manure from my compost heap over the field, *after the first plowing, and before cultivating or harrowing*. This is thoroughly worked under (and but just under), by a small one-horse plow, driven at right angles with the furrows, after which I follow with the cultivator, aiming to have everything as thoroughly fined up as possible. If time presses, I dispense with the small plow, and depend wholly on the cultivator and harrow to get my manure under the surface. After the manure is well worked under, the hills or drills are marked off by dragging a chain over the surface, the first line being made straight by setting up two poles ahead, and keeping them in line while walking; afterward the lines can be kept conveniently straight by carrying a pole of the same length as the distance desired between the hills, and using it occasionally as a guide. After the field is thus chained out in one direction, it is crossed in the opposite direction. The hills are marked out by the crossing of the lines made by the chain. If the surface is free from large rocks, the hills can be marked out by running two sets of furrows, the hills being made where they cross each other at right angles.

In the hills I work in my manure, avoiding all stable dung, or any animal manure, as this is liable to contain seed, and to one who raises squashes for seed purposes, this is quite a serious objection, for, in fact, I have found it almost impossible to keep squashes pure, where animal manure is used in the hill. I manure in the hill, or drill, with the most highly concentrated manures to be procured, such as guano, superphosphate of lime, or fish guano. The reason for using highly stimulating manure in the hill is, to give the plants a quick start when young, that they may grow beyond injury from the ravages of the striped bug.

There is danger in using highly concentrated manures in the hill, that the roots of the young plants be destroyed —"burned" is the farmer's phrase; to prevent this, they should be most thoroughly stirred in with the soil. My practice is, to take such manure in a wooden bucket, and passing from hill to hill, scatter, if phosphates, as much as I can take up in a half closed hand; if Peruvian guano, about half as much, over a circle of about two feet in diameter. A man follows immediately after with a six-tined fork; he is directed to turn it just under the surface, and then draw his fork across the hill three times, and again three times at right angles with the first direction, ending with planting the fork in the middle of the hill, and giving it a twist around. I am thus particular in my directions, because day laborers seldom realize the corrosive effects of these highly concentrated fertilizers. After my man, a boy follows to plant the seed; he sweeps a circle with his finger around each hill, as he finishes planting.

After the vines have got so far along as to show their runners, I top dress the surface with hen manure, or some of the special manures above mentioned, and immediately follow with the cultivator.

It will be perceived that my system of manuring is based upon the theory that vines prefer their food near the surface of the ground. I draw this inference from the fact, that vines are great lovers of heat, being quite sensitive to changes of temperature, and also from tracing roots, and finding under the old system of deep manuring, that they would, at first starting, run but an inch or two below the surface of the earth, when they would spread out horizontally, and stretch on for some feet at a very uniform distance below the surface. Again, I find my crops very satisfactory under this system of manuring, and for the past four years have cultivated all my crop (four to seven acres annually), on this plan. My friends will note that I reduce my manure very fine, and mix it very thoroughly with the soil. My soil is a strong loam.

PREPARING THE HILLS.

The system almost universally advised and pursued in preparing the hills for planting, is to throw out the earth from within a circle of from two and a half to four feet in diameter, and from six inches to a foot in depth, oftentimes quarrying out rocks and digging into the hard-pan to get the standard depth. Then fill in with manure, and cover this with earth, raising a low mound in the form of a truncated cone about six inches above the surface. On this mound the seed are planted. Where the land is freshly turned sod, the hills are usually made by cutting a hole of the usual diameter in the sod with a sharp spade or axe. In my own practice, I have given up this method for years. The plan of excavating a hole, and putting in it all, or about all, the manure for the crop, appears to be founded on the theory that the roots will confine themselves to the area—an idea entirely erroneous, as we have already shown. Quarrying into the hard-pan and putting manure down to such cold depths, is inviting the vine to violate its instinctive love of heat. Again, this system involves a great deal of labor, particularly when sod land is planted, and on these latter the pieces of sod taken out of the hills remain nuisances over the surface of the field, either clogging the cultivator, or being knocked against the young vines. Let any farmer try the plan of preparing his hills as I have detailed above, and I will guarantee that he will not again return to the present system. If barn manure is to be used in the hills, let them be made saucer shape, broad and shallow. In preparing freshly broken sod, I find Share's harrow an excellent implement, as it will pare down the sod to an inch in thickness, and make the soil as easy to be worked as old ground.

HOW FAR APART SHOULD WE HAVE THE HILLS, AND HOW MANY VINES SHALL WE LEAVE IN THE HILL?

The great error among farmers is to make these hills too near together, and leave too many vines in each hill. A very common distance for Marrow squashes is six feet apart each way, three or four vines being left in each hill.

A little figuring will show the bad policy of the practice. When a Marrow squash vine grows alone—and it oftentimes happens that one comes up among other crops on the farm—it will mature as many as three squashes, and at times half a dozen or more. Squashes so grown are almost always fine types of the particular variety. Now, on the contrary, when the hills are six feet apart, with three or four vines to a hill, the vines will not average *one* squash to each. I have been amused to receive the estimates of farmers of the number of squashes to the vine on the heaviest crop of Marrows they ever saw. As often as not the reply would be "three to the vine." Now an acre of ground planted 6 × 6 will have about 1200 hills to the acre; four vines to the hill would be 4800 vines to the acre. The present variety of Autumnal Marrow squashes as now grown, will average above seven pounds to the squash; if the vines produced on an average *one* squash apiece, we should then have 33,600 lbs., or over seventeen tons to the acre! Whereas the largest crop on record, as far as I am aware, of this variety of Marrow is less than eleven tons to the acre. From such figures the conclusion stands out with emphasis, that a system that, taking the average of crops, does not give over one squash to *two* vines, is unnatural, unfarmer-like, and unprofitable.

The shortest distance, where the hill system of planting is pursued, should not be less than 8 feet each way for Boston Marrow squash and other running varieties, with the exception of the Hubbard, Turban, and Yokohama, which are ranker growers, and should not be planted nearer than nine or ten feet each way. The hills for the Mammoth varieties should be twelve or more feet apart each way. At these distances apart, two plants in each hill, (the vines being thinned down to that number when the runners begin to start), will be found sufficient to well cover the ground. Were it not for danger from the borer, I would never leave more than one vine to a hill,—putting the hills in each case proportionally nearer. One of the finest crops of Turban squashes I ever raised, a crop that took the county premium for yield that year, was raised with but one vine to the hill, and the crop that took our county premium the year previous was grown with two vines to the hill. This brings us to the discussion of the Drill versus Hill system of planting. On the supposition that the great error in growing squashes has been to crowd the roots too much together below ground, while the vines were crowded too much together above ground, I have advocated, and to some extent practised, the Drill system of planting—having each vine entirely by itself, and distributing them evenly over the ground. Assuming that 10×10 or 100 square feet is sufficient room for the plant, on the Drill system, I allow 7×7 or about 50 feet for one plant. In planting on this system, the field is marked out as if for hills, the lines crossing each other every seven feet. In planting in drills I put three seeds along in the line, and when the plants begin to show runners, thin to one plant. By the drill system, in addition to the advantages above claimed, I think that the crop is more uniform in size, and the squashes are better proportioned in their forms than under the hill system. The vines being in a row, instead of a circle, the cultivator can be carried nearer to them. Most of my land is very uneven, otherwise I should always plant in drills in preference to hills.

PLANTING THE SEED.

The quantity of seed per acre for the Marrow and Hubbard varieties is set by practical farmers at two and a half pounds. This allows for liberal planting with a good surplus for after use, should cold or wet weather rot the seed, or insects destroy the plants that first appear. Four seeds in the hill and three in the drill is sufficient. The seed should not be put in, in the latitude of Boston, *earlier* than the 10th of May, and may be safely sown in ordinary seasons as late as the first of June, and success is sometimes attained with seed planted on rich, warm land as late as the twentieth of June. A part and sometimes all of the seed planted as early as the 10th of May will rot in the ground; yet to get the vines along early, and thus enable them to survive the attacks of the squash bugs, farmers oftentimes take this risk. If, after a cold, wet spell, the planter mistrusts the seed have rotted in the ground, let him scratch away the earth carefully with his fingers (it is infinitely easier to put a seed under than to find it again!), and if the seed is rotten, it will readily show it when pressed between the thumb and finger.

Seed may be planted either by using the hoe, (dropping the seed, and covering with the hoe,) or each one may be thrust into the ground with the thumb and finger. If the attempt is made to push the seed under by the finger alone, it is frequently left too near the surface, as the finger is very apt to slip by it unawares. If squirrels or field mice abound, it will be found safer to plant with the hoe, as the little rascals appear to have a rare faculty for smelling out the very spot where the seeds lie when thrust under by the finger. I have known them to begin at one end of a field and pass from hill to hill in a straight line across the field, digging out every seed with unerring accuracy. Seed opened with a knife and rubbed with arsenic or strichnine and scattered in the paths will generally check them. Two inches is ample depth in any soil, and early in the spring, or in a rather wet or heavy soil, the seed had better not be planted more than from an inch to an inch and a half in depth.

Seed planted on upturned sod will vegetate sooner and come up with larger rudimentary leaves than that planted in rich, old ground; I presume that this is because sod land lies lighter and is better drained and, consequently, warmer than old ground. If, when the rudimentary leaves appear, the seed shell adheres to either leaf, it will do no harm, but if it confines both leaves together, it should be removed, if it can be done without injury. If a seed pushes but a single rudimentary leaf above the surface, the plant rarely, if ever, comes to anything. If these rudimentary leaves continue to increase in size, but no leaf shows itself springing from between them, the plant will come to nothing. If the young plants come with a yellow color, it proves that the season is too cold for them; if, on the other hand, they assume a very dark, dull green color, it is usually because the manure with which the young rootlets are in contact is too strong for them; it is good policy, when the manure proves too strong, to carefully remove some of the earth around the plants with the finger, and with the finger stir in a little fresh earth.

If, as at times will happen, some hills are entirely destitute of plants, it is far better to plant them with seed than to transplant surplus vines from other hills; true, such vines sometimes root at once, but if checked in their growth by transplanting, they rarely amount to anything in the end.

This is one of the great conditions of success in squash culture, to *have the vines start well and make a rapid growth without a check.* Experience has frequently proved that late planted vines will oftentimes ripen their crops as early, and usually bear heavier crops, than those planted two or three weeks sooner.

HILL CULTURE AND LEVEL CULTURE.

After the plants appear, it is customary to draw earth around them; this is a good practice as far as it tends to keep them from being broken off by the winds. It is also an almost universal custom to draw up the earth into a mound of two or three feet in diameter, gradually increasing the height of it with each hoeing until it is six inches or more above the level of the field. I consider the labor entirely useless, to say the least, and have confined my own practice for several years past to level culture, making no hills, and drawing just earth enough home to each plant to keep it from being swayed, and thus injured by the wind.

HOEING AND CULTIVATING.

About as soon as the plants show themselves above the surface, the Cultivator should be set running. If the hills have been made equi-distant each way, the surface can be cultivated close home to them on every side, leaving but little work for the hoe. In no department of farming is the superiority of the Cultivator over the common hand-hoe brought out in stronger contrast, than in working the large open areas between squash hills. I would rather have the work done by a one-horse Cultivator with a boy to direct the horse and a man to hold the implement, than have the services of twenty men with hand hoes; for not only would the surface be gone over in equal time, but the ground be more deeply and more thoroughly stirred, and the weeds be better shaken up and turned under than would be possible with hoe culture. The cultivator should be used as often as the weeds start, and whenever the surface appears hard, the object being two-fold, to eradicate weeds and keep the surface light and mellow. If witch grass abounds, the Cultivator must be freely used, particularly when the surface is hot and dry, that the vitality of the freshly torn roots may be destroyed. It is not well to leave the soil unstirred until weeds have attained to some size, as such are very apt to re-root. If the Cultivator is used while the weeds are small, it can be spread open to its utmost capacity. It is always well to have one course of the Cultivator half overlap the preceding course.

The last, and one of the most critical, periods when the Cultivator is needed, is just previous to the pushing out of the runners over the surface of the field. The vines are then growing rapidly, (I have found that the large varieties, by actual measurement, grow as much as fourteen inches in forty-eight hours), and if special care is not exercised, the runners will push so far as to prevent the final use of the Cultivator. The result will be a very weedy field the remainder of the season. I have sometimes practised, when caught in this way, breaking the hold of the tendrils and turning aside with the hand such runners as had got so far from the hills as to be in the way of the Cultivator; but I have observed that where the tendrils are broken from whatever they have naturally clung to, as often as not the vines are injured so much by the wind that they yield little or nothing; they are so twisted that they are often completely inverted; and though the leaf stalks are true to their instincts, and bring themselves perpendicular to the surface, yet in doing so, the curve they make, passing under the vine, lifts it a little above the surface, too far for the joint roots to strike into the earth to hold the plant in place and nourish it. It is a bad plan ever to break the hold of the tendrils, and as a general rule better allow the large weeds that appear towards the close of the season to remain, than to pull them up and tear them out from among the vines. If the weeds are to be removed, better cut them off close to the surface and leave them. A squash crop will foul the land at the very best, and let no one plant to squashes with the idea that the frequent cultivation allowed early in the season will tend to improve a piece of ground already foul with weeds; for young weeds will spring up as soon as the spread of the vines prevents the farther use of the Cultivator, and when the leaves begin to thin out, at the close of the season, under the stimulant of the sun and air, these soon become mammoths in the rich soil. When we consider that climbing appears to be natural to the squash vine, the injury caused by breaking the hold of the tendrils, and by the moving about among the thick net work of vines to do this, in connection with the fact that at best it is next to impossible to keep the ground in clean condition, I question whether, as a general rule, it is not better to allow these late and large weeds to remain untouched, and leave the clearing of the ground to the crop of the next year.

When the area of ground is small, and very clean culture is desirable, I would advise the driving of a few stakes among the vines to give the runners a hold when they first push out. It is not necessary that these stakes should protrude more than one or two inches above the surface.

Many old farmers lay down the rule that no one shall set foot on the squash patch after the vines meet between the rows. This is a good general rule, for most men tread among vines as ruthlessly as though passing among wire cables, crushing them under foot with perfect impunity. I don't think I ever saw a farmer pass among even his own vines with what I should call proper care. If necessary to pass among vines, carry a short stick in one hand to lift the leaves to see where the foot is to rest before planting it.

SQUASHES WITH OTHER CROPS.

In the vicinity of large cities, where land, manure, and labor are costly—and much of the market gardening in the vicinity of Boston, New York and Philadelphia is on land worth from $500 to $1,000 an acre—farmers usually grow their squashes in connection with other crops. These are oftentimes Peas and early Cabbages. If early Peas or Cabbages are planted in rows three feet apart, by omitting every third row, and planting this to squashes at the usual time, the crops will not interfere with each other, as the squashes do not push their runners till July, after the pea crop has been marketed. With Cabbage, the third row may be omitted, or every third plant in the third row; this will give the squashes 9×9. It will be seen that squashes can be raised only with the earliest varieties of Cabbage, such as Early Wakefield, Early Oxheart, Early York, Little Pixie, Burnels, King of Dwarfs, that have been started in a hot bed. The plan practised occasionally of growing squashes among corn, I consider a bad one. It is very common in the country to plant at the second hoeing a couple of seed of the Yellow Field Pumpkins in every third or fourth hill, and the yield is usually satisfactory to the farmer; though if a field was divided in two, and an accurate account kept of the income from each half, I am inclined to believe that it would be found that what was gained in pumpkin was more than lost in corn. Squashes are more delicate in their habits than the hardy, rough vined pumpkin, and the result of attempting to grow them with corn is usually a small crop of inferior specimens.

SETTING OF THE FRUIT.

Soon after the runners have put forth, blossom buds will begin to appear at the junction of the leaf-stalks with the vine. As the buds develop, the stems will develop also, until the latter grow a foot or more long, a little longer than the leaf-stalks. The blossom now opens, and we have a large yellow flower, several inches in diameter, with a powerful and rich fragrance, very similar to that of a magnolia. This flower has at the center a yellow cylinder, about an inch in length, covered with fine yellow pollen. I find that many persons look for their squashes from this class of flowers. Squash vines have the sexes distinct in each flower, being what botanists call monœcious. These are the male flowers, and from their structure can never produce squashes; their office is wholly to supply pollen to fertilize the pistillate or female flowers. The first pistillate or female blossom rarely appears nearer the root than the seventeenth leaf, or farther than the twenty-third. Instead of having a long stem to support it, this flower opens close down to the juncture of the leaf-stalk with the vine. It has a small globular formation beneath it, which is the embryo of the future squash. If the structure of the center of the blossom is examined, it will be found to differ from the tall, male flower, in having the central cylinder divided at the top into several parts, usually four, sometimes six in number. These are what botanists call the pistils, and it is necessary that the fine yellow dust of the male flower should touch these, to fertilize them, that seed may be produced, and consequently a squash grow—for the *primary* reason why a squash grows, is, to protect and afford nutriment to the seed, the use of it as food being a secondary matter. This may be proved by so confining a blossom, that no pollen can get access to it, when the blossom will usually wilt, and the embryo squash turn yellow and decay. If the female flower be broken off from the embryo squash before the flower has come to full maturity, the squash will decay. These female blossoms are so covered and hidden by the tall leaves, that it is evident that the fertilizing pollen must be conveyed to them by the bees, to whom the squash field appears to be a rich harvest field. All of the crossing or mixing of squashes is caused by the pollen from the male flowers of one variety being carried by the bees to the female flowers of another variety. SQUASHES ARE CROSSED OR MIXED IN THEIR SEED, AND NOT IN THE FRUIT. Many cultivators are in error on this point; they have the very common illustration of the crossing of different varieties of corn in their mind, where the mixture of the varieties is at once apparent to the eye, and infer from this, that the mixture between different varieties of squashes should make itself visible to the eye *the same season it occurs.* A moment's reflection will correct this; *the crossing of the first season is always in the seed,* and for this reason we see it in the corn the first season, as the seed is immediately visible to the eye, while the various colors of the different varieties also aid us in the matter. With squashes the crossing is likewise in the seed, and hence can not be seen in them, until the seeds are planted, when the yield will show the impurity of their blood. But, though the crossing can not be seen in the squashes themselves the first season, yet, if one of the varieties planted near each other, has seed having the peculiar, thick, salmon-colored coating, so characteristic of some of the South American varieties, this indication of admixture may be detected by the eye the first season. The parallelism between the crossing of squashes and corn may be carried further, for it is oftentimes true with corn as with squashes, that there is a mixing of varieties, of which no indication can be detected in the seed by the eye the first season, which a second season will develop—what was before an eight-rowed variety, into a ten or twelve-rowed sort, or dark kernels may be replaced with white ones, and by numerous similar freaks, bring to light an admixture of varieties.

It is of considerable practical importance, that the law of admixture should be clearly understood, that the risk, incidental to planting seed from squashes that *look* pure, should be generally known ; for it will be seen from what I have written, that seed taken from squashes that externally are perfect types of their kinds, may yield a patch, where every one may show marks of impurity. Again, no matter how many varieties are planted together, no crossing from the result of that planting will be seen in the *external shape, color, or appearance* of the crop *the same* season.

To have squash seed pure, the squashes from which they are taken, must have been grown isolated, and this not only one season, but for a succession of seasons. Should several varieties of squashes be grown together, and it be desirable to keep one variety pure, it can be done by preventing any male flowers of the other varieties from maturing—no easy job, as those who have tried it know. The product of any particular blossom may be kept pure under such circumstances by covering with fine muslin, removing it only to fertilize with pollen from a male flower of its own vine.

The location of the female blossom, in a measure covered by the leaves, and low down, but little affected by the wind, would render it probable that it depends for fertilization on the bees, rather than on the wind; and the fact (as a friend who has tested it, informs me) that if only a high fence intervenes between two varieties, the admixture between them is comparatively small, corroborates this theory. To preserve the degree of purity that is necessary in raising different varieties, requires planting at distances apart varying with the natural aspect of the country; a level tract requires longer distances than would be necessary in an undulating country, and a space intervening abounding in flowers is a better protection than an equal distance where flowers are less numerous. The object is to get the pollen removed from the thighs or bodies of the bees, or have it covered by the pollen of other flowers, before they can pass from a field of one variety of squash to that of another. My own practice is, to secure the planting of one continuous district of country with the same variety of squash, by giving to farmers, whose lands are near my own, my stock seed for their own planting. Even with this precaution matters will have to be looked after, lest after all promise to the contrary, greed can not master moral courage sufficiently, to induce them to pull up the transient vines that spring up from the manure among cabbages or potatoes. Old farmers will profess, from the appearance of the calyx end, to classify squashes as male or female; this is all nonsense, for, as will be inferred from what has been stated, every seed from every squash contains the two sexes in itself, in its capacity to produce both male and female flowers.

Squash fields usually make about three settings of fruit. I do not mean by this that *each* vine makes three settings, but that this is usually true of a field as a whole. It often happens, that most of one of their settings, usually the second, turn yellow and rot, after many of the squashes reach the size of goose eggs. This is very apt to take place, should there be a cold, wet spell just after they have set. Sometimes all three of the settings will grow, and then stories of great crops will be heard of in the squash districts. When a young Hubbard squash is making a fine growth, it will have a shining green appearance, as though just varnished, If the appearance of the squash changes to a dull green color, the days of that squash are numbered; it will soon shrivel and decay.

PINCHING VINES.

I have seen a vine perfect the growth of a squash 20 lbs. in weight, though the vine was cut off within a foot of the squash when it had reached the size of an orange, and another squash of about the same size was also matured on the same vine, about four feet nearer the root. The vine was highly manured, and grew on very deep and rather moist muck and loam. I can not yet determine the laws which govern the art of pruning vines. I have had some, the young squashes of which appeared to do finely after the extremities of the runners were nipped at near the close of the season, and others, where the young squashes turned yellow and died, under, seemingly, precisely the same circumstance. I am inclined to think, that it is not well to pinch off the ends of the vines before the young squashes have attained to the size of a large orange. How far a crop of squashes might be increased by the nipping of the vines, or a pruning of the roots, is a problem yet to be settled. The use of the cultivator just before the vines spread, must do much in the way of root-pruning the vines.

THE RIPENING AND GATHERING OF THE CROP.

In seasons, in which the early part of summer is cold, farmers sometimes get almost discouraged with the small number of squashes that set, and the slow growth of such as do form, but a few hot weeks entirely change the aspect of affairs.

When we have good corn weather, it takes but a few weeks to mature a squash. I have known instances when the first fruit set was completely destroyed by a hail storm, which occurred late in September, and yet a fine crop of squashes was gathered from the vines. When June and July are colder than usual, farmers will often come out from an examination of their squash patch with a significant shake of the head, yet I have never known a season, in which cold or wet prevented the growing of a fair crop of squashes on land selected with judgment, well manured, and taken care of. The degree of ripening to which the crop attains, will be affected by a cold and wet season, but the chances of a crop are equally good with a season wetter and consequently colder than usual, as with a season hotter and dryer than ordinary, for, in addition to the check to their development caused by a drought, the borer and bugs are more numerous and more active in a very dry season than during a very wet one.

Ripening is indicated in the soft or fleshy stemmed squashes, such as the Hubbard, Marrow, and Turban, by the drying of the stem, and a dead, punk-like appearance which they assume. The leaves near the root gradually turn yellow and dry up, and the squashes themselves change color; the Hubbard assuming a duller, more russet color, and the Marrow and Turban sorts a deeper orange. The skin of the Marrow and Turban will now offer more resistance to the thumb-nail, while the Hubbard will begin to put on a shell, which will be first detected near the stem end. It is a singular fact, that the shell of the Hubbard squash usually begins to form on the under side—the part towards the ground. When this stage is reached, squashes can be safely cut for storage.

At some seasons, a large portion of the crop, and, at most seasons, a small portion of the crop, just before ripening, are affected by a blight, which turns the leaves black near the hills, when they die down, and all the signs of early maturity are presented to the inexperienced eye. When the process of ripening goes on naturally, the exposure to the sun's rays, after the leaves have died, does no harm, but promotes the full maturing of the squash; but when squashes become exposed before the natural time, by the blighting of the leaves, they are, particularly if of the Hubbard variety, apt to be "sun scalt," as the term is, by which is meant a bleaching, or whitening of the part most exposed to the sun. Such squashes rarely form shells, and, if badly scalded, are apt to rot at the part affected. In cutting squashes from the vines, a large and sharp knife is needed. There are two ways to cut squashes from the vines; one is, to cut the vine, leaving a small piece attached to the stem. By so doing, the stem does not dry up so readily, and as large stems, when green, will weigh as much as a quarter of a pound, if squashes are to be sold soon after gathering, this will give quite an addition to their weight. Narrow, selfish men sometimes cut their squashes this way.

The usual way is, to cut the stem from the vine. When first cut, more or less sap will run out in a stream from the hollow stem, though the squash may be fully ripe.

A CRITICAL PERIOD.

What shall be done with the squashes after they are cut from the vines? The stems need a little exposure to the sun to scar them, and the earth, which adheres to those grown on low land, needs to be dried, that it may be rubbed off before the squashes are stored. A good way to accomplish this, is, to let the squash remain where it is cut, provided the leaves do not shade it, care being taken to give it a turn, to bring the under side up to the sun.

If there is danger from frost, it is better to gather them

together at convenient distances, that they may be more readily protected. The interval between the cutting of squashes and the storing of them is a critical period, as oftentimes the keeping of the crop depends upon the course then taken. There is a pernicious practice, quite prevalent, of placing them in piles as high as can be made, without their rolling off. Should frost threaten, this, of course, is necessary in order that the mass may be the more readily covered with vines to protect them; but when so piled, as soon as danger from frost is over, they should at once be taken down, so that all may be exposed as much as possible to the sun and air. Farmers, in handling squashes at this period, are apt to lose sight of one important fact, viz.: that when a squash is cut from the vine, its vitality is impaired, and it has no longer such power to resist the effects of atmospheric changes as it had previous to the separation. I say its vitality is "impaired," for the fact that the seed continues to fill out for a month or two after the squashes are gathered and stored, proves that there is a degree of vitality, however feeble, yet remaining in the squash after separation from the vine. The fact that sap exudes and gradually thickens into tears, or, at times, runs in a stream from the stems when cut, no matter how ripe a soft stemmed squash may appear to be, seems to prove that some vital function of the sap vessels has been disturbed; while the greater readiness with which such squashes decay, carries us beyond theory to the *fact* of a diminished vitality. I have known the lower layer of a lot of Marrow squashes in the field, to be found rotten through and through on removal— and this when there had been no frost to injure them—the result being due wholly to the dampness of the ground, during a rainy interval, acting on a diminished vitality.

I have known instances in which lots of Marrow squashes that had never been touched by frost, and were perfectly sound when stored, were suddenly covered with spots of black rot, soon after they were put into a dry apartment. These lots had been exposed in the field in piles during a series of days of cold rain. The practical lesson to be drawn from such facts is, that squashes should never be left in the fields exposed to cold rains after cutting.

After the stems have had the sun a couple of days to dry and sear them, and even before, if cold, wet storms threaten, the squashes should be piled with great care on spring wagons, and taken from the field. The rule should be laid down as invariable, that no squash shall be dropped in any stage of its progress, from the field to the market; they should always be *laid* down.

THE STORING OF THE CROP.

Squashes are usually at their lowest price in the fall of the year, after the crop has been gathered, and before the first severe frosts. The crop being bulky, and requiring dry storage, farmers are intent on getting it to market before cold weather sets in. After the first severe freezing weather, the crop is usually held at a higher figure, as the surplus not intended for storage has been disposed of. In the immediate vicinity of the large cities of the North, a large proportion of the crop is stored in buildings known as "squash-houses," to be marketed during the winter and spring months. These buildings are oftentimes old dwelling-houses, school-houses, or ware-houses, removed from their original locations to the farm, and then put to this secondary use. I present a vertical section of my own squash-house, by which the general features of all of them can be seen at a glance.

In dimensions, the building is about 24×35 feet, with a height of 10 feet to the plates. It is divided into three rows of bins, which are separated from each other and the sides of the building by aisles, (*A, A, A,*) about 26 inches in width, a distance which admits of the easy handling of a bushel basket, or barrel. The bins, (*B, B, B,*) are about 5 feet wide, 26 inches high, and 5½ feet long. The uprights, which support the series of bins, are small joists, 2×4 inches, with cross-ties of inch or inch and a quarter board sunk into them, on which the several platforms are laid. These uprights are the length of the bins apart, viz.: 5½ feet. At the edges of the bins, boards, 6 inches

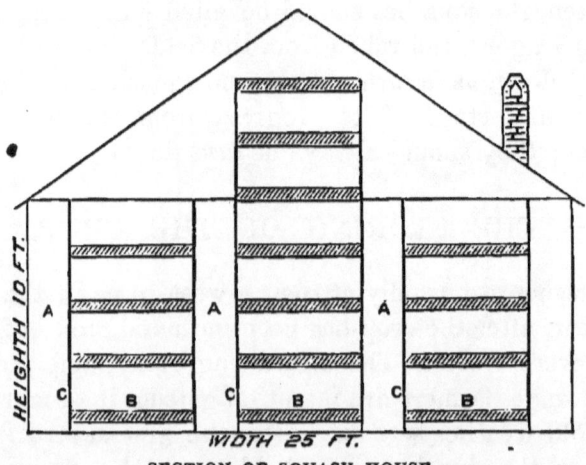

SECTION OF SQUASH-HOUSE.

wide, are laid, to prevent the squashes from rolling out. These boards should be planed on the inner, upper edge, that they may not cut into the squashes that lean upon them. The series of floors are made of strips of board, from four to six inches wide, nailed about half an inch apart, to allow a circulation of air. It is well to have the lower floor a sufficient distance from the floor of the squash-house, to permit a cat to go under. The cellar wall should be carried close up to the floor, by filling in front of the timbers with brick, or small stones and mortar; this will prevent rats from working through. As the building is designed to support much weight, it should be strongly braced by timbers crossing from plate timber to plate timber, to prevent spreading, while the cross-timbers, in the cellar, require props of masonry, or joist. To economize in fuel, on the two coldest sides, my squash-house is double plastered, and has double windows all around; some have inner wooden shutters to each window, which are kept up during cold weather, both day and night, only as much light being admitted, at times, as may be necessary, while attending to work. The roof has five sliding windows, which assist in ventilation and give light to the upper part of the building, that otherwise would be quite dark when filled with squashes. The stove is at one of the coldest corners, with a funnel passing across to a chimney at the opposite corner. A building of the above proportions will hold about one ton of Hubbard squashes to two bins, and by careful and close stowage in all available room, it can be made to hold about sixty tons.

There is an advantage in having a low, wide building rather than a high and narrow one, as a greater portion of it is accessible from the floor, it is less exposed to cold winds, and the heat is more evenly distributed. In a high building, the heat in the upper portion is apt to be excessive.

The squashes should be brought to the squash-house in a dry condition, and be stored before dew falls. The stems being yet green, the squashes should be so piled as to bring these to the outside as much as possible. In placing the squashes on the shelves, put the largest ones on the bottom, giving them all a slant in one direction; they will thus pack better, and the uniformity will be agreeable to the eye. From the beginning of the storing, every window and door should be kept open during fair weather, and a fire at the same time will help in the drying of the stems. Should there come a damp time of one day or more, by all means start the fire. The stems will be apt to mould some, and the air of the building have a disagreeable smell if they decay, though a little moulding may always be expected. In about three weeks from the time of storing, the stems will be dry. In handling the squashes, I need hardly reiterate the caution of care. My practice is to form a string of boys, from the wagon to the shelves, and the squashes are tossed from one to another, with the caution to handle them like eggs. Boys well trained will not drop more than one squash to the ton, and I have known my boys to pass several tons without dropping a single squash.

CARE DURING THE WINTER.

If the squash-house has been built with reference to warmth, when once filled with squashes, it is surprising with what little fire frost can be kept out. The mass of squashes are, in themselves, a great store-house of heat, and with inside shutters for the coldest weather, the building is frost proof, with a small outlay of fuel.

In my own building, capable of storing sixty tons or more, I have a salamander stove of capacity sufficient to hold two hods of coal. In ordinary winter weather two hods of fresh, and a hod of sifted coal for night use, will last about twenty-four hours. To keep the fire over night, I leave the cover off about half an inch, and, if very windy, also put up the door in front within half an inch of closed. When I first attempted to keep squashes during the winter in very cold weather, I frequently sat up till midnight, and then retired with much anxiety, lest Jack Frost should steal a march on me before morning; but from experience I find that a salamander can be as well regulated and as readily controlled as a Magee stove, while the greater length of funnel that can be used with them, by reason of their superior draft, is a decided advantage.

No one can keep squashes to the best advantage, until he has fully learned to so control his fire as to keep the temperature near the freezing point, and yet not endanger the squashes. From a want of this knowledge, almost all squash-houses are kept at too high a temperature, and, as a consequence, the squashes lose in weight and quality, and, if they are Hubbards, in appearance also, losing their fine dark green color, and becoming of a reddish, rusty hue. The best temperature is as low as forty degrees. After squashes are stored, the great desiderata are a low temperature and a dry air. Should the weather be mild in the course of the winter, never be tempted to open the windows *unless the air is dry,*—a very rare thing in winter, as, on most mild winter days, the air is loaded with moisture. If it is desirable to air the squash-house, select a dry day when not very cold, start up the fire and open the windows at the roof. Squashes that were grown in a wet season, will rot most in winter, and *vice versa*. Other things equal, the keeping of squashes depends greatly on the hygrometric state of the air—in other words, the dryer the air the better they will keep. This is the reason squashes keep better in a squash-house than in a cellar—the house is no warmer than a cellar, but the air is dryer. In dry, sandy cellars, by the aid of a fire, they can be kept about as well as in a squash-house. Squashes in dry cellars will usually keep very well until January, and sometimes to the first of February, particularly if the damp, external air can be kept from them. Several years ago I lost not far from twenty-five tons of squashes in about ten days, as I now believe, from having admitted the warm, damp air of a January thaw into the cellar. After squashes are stored, the less they can be handled the better; and in cellars, it is oftentimes better to let a few rot than to overhaul squashes late in the season with reference to culling out the rotten ones, for, after such overhauling, they usually decay faster than before. Cellar-kept squashes have some advantages over those kept in a squash-house; they keep their original rich green color, lose but little or none in weight, and are of better quality. They have the two disadvantages of not keeping as long, and perishing very soon when sent late to market. This latter fact is now generally known to dealers, and they hesitate to purchase cellar-kept squashes late in winter. The winter of 1866–7 will be a memorable one among the squash men of Massachusetts. Squashes being remarkably plenty and cheap in the fall, every squash-house in the vicinity of Boston was filled to overflowing. As the season advanced, squashes began to show a remarkable tendency to rot, and the result was that, in many cases, as large a proportion as four-fifths of the crop rotted before spring opened. The summer previous had been unusually wet and cold.

If apples, squashes, or any other fruits are gathered ripe, the next step is to decay; but if they are not fully ripe, they have this intermediate step to take before decaying. Heat is an agent in promoting progress in each of these steps; hence, the less heat above a freezing temperature in which squashes can be kept, other conditions being equal, the longer they will keep.

The very small squashes which are usually given to stock as soon as gathered, are among the very best for keeping, provided they are stored in the warmest part of the building. Late in spring they are salable at a high figure for cooking purposes. Out of about five hundred pounds of such squashes stored so near my salamander that the outer tier cooked with the heat, I found but about ten pounds of defective squash when I overhauled them for the first time, near April. Squashes planted about the first of June will usually keep better than those planted earlier, on the same principle that the Roxbury Russet, and Baldwin, keep better than the Porter, or Sweet Bough apple, the former not being ripe when gathered from the tree. The order in nature is that fruit should ripen before it decays.

MARKETING THE CROP.

Squashes are sold by the piece, by the pound, and by the barrel. Sales by the piece are unknown in the Eastern States, as far as my knowledge extends. In the markets of New England, after the summer squashes, of which there is but a limited demand, the Marrow and Turban are brought to market, and, before frosty weather sets in, they are sold mostly by the ton to large dealers. Late in the fall the Hubbards begin to come to market, for if sold just after gathering, they are rather forced on the market, the Marrow and Turban being usually recognized as the squashes for fall use. During the winter, the supply from the squash-houses around Boston is mostly brought to market in barrels, and sold by the barrel without weighing. This is poor practice, as there is often a number of pounds difference made by the thickness of the squash, its size, the packing, and the size of the barrel. Such a system of marketing is apt to tempt to petty trickery.

A greater or less proportion of stored squashes will decay under the most favorable circumstances. It is the policy of the squash grower to lose as little as possible in this way, and the custom of the markets of Boston usually allows a little latitude in this matter. Hence, particularly as the season advances, one or more squashes that have small rotten spots on them, are often packed in a barrel. The Hubbard is a very deceiving squash; it may be entirely rotten inside, and yet, to inexperienced eyes, appear perfectly sound without. If the outside has white mould spots, looking like some of the concentric mosses, the squash is usually sound underneath the shell; but if these mould spots are greenish or yellow, it is usually soft rotten in a spot just beneath them. If the shell at either end, (and the Hubbard usually begins to decay at the ends), has a watery look outside, the squash is usually considerably decayed underneath. If the Hubbard is very light, it has usually the dry rot inside; if remarkably heavy, it is usually water-soaken and worthless within. If a squash, on being cut, proves to be water-soaken, a close examination will usually show some small opening, where, during some stage of its growth, the external air found entrance.

FROST-BITTEN SQUASHES.

With the utmost care, squashes will at times get frost-bitten. The Marrows and Turbans show this by turning a darker orange color on the part frozen. If as much as one-half of the squash has been frozen, it is frozen through its thickness, and will very certainly soon decay, and the best disposition to make of it is, to keep it at about freezing point in an ice-house, until fed to stock. If less than half has been frozen, before the sun shines on it turn the frozen surface under, and keep out the light as much as possible; this will take out the frost and save it, if any remedy will, though a frozen squash is always unreliable property. Some years since, I had a load of Marrow squashes brought me, which had been stored in a barn during a cold spell, and the outer tiers had been frost-bitten. I separated the badly frost-bitten ones, putting them, frozen side down, in a dark cellar on the damp earth, and stored such as showed no signs of injury on the shelves. In a few days, no sign of frost could be seen on those stored in the cellar, and they kept apparently as well as though they had never been injured, while those stored on the shelves soon rotted badly. The Hubbard squash is not as much injured by frost as are the Marrow and Turban; if it has a shell on it, the result will usually be the production of a dry rot under the shell as far as the frost extended, and no further. I have cut squashes in February that had been frozen in November, over an area of about five inches square, and found all the injury done limited to this space.

MARKET PRICES OF SQUASHES.

Within the past six years, Marrow squashes have varied in price in the markets of New England from $10 to $40 per ton; these variations are caused, for the most part, by the quantity brought to market, for, though equal areas may be planted, there may be all this difference, owing to the greater prevalence of insects one season over another. The average price of Marrow squashes for the past six years has been about twenty-five dollars a ton.

The extremes of prices of the Turban and Hubbard during the same period have been from $20 to $50; the average having been nearly thirty-four dollars.

Previous to the war, the Marrow ruled in the market at from $15 to $20 per ton, and the Hubbard at from $20 to $25. These prices are the market rates just after the crop is gathered. As the season advances, prices rise to 50, 60, 70, 80, 90 and 100 dollars per ton, and occasional lots kept late into the spring, and sold by the barrel, have brought as high as $140 per ton. The last four tons I sold the past season brought me $400; yet so remarkably poorly did the crop keep the past winter, that the profit would have been equally as great, had I sold at $25 per ton in the fall.

The markets of New York and of the large Southern cities are, as yet, but poorly supplied with the Hubbard squash during the winter season. I can think of no investment in agricultural products that would pay better than the judicious handling of a couple of hundred tons of Hubbard squashes in New York or Philadelphia during the winter months.

Squash farming, on lands pushed well out into the ocean, have some advantages over inland farming. Neither the cabbage, or turnip fly, the pea bug, squash bug, or other destructive insect is nearly as prevalent in such sections as just back from the coast, while the temperature is three or four degrees higher late in the fall, which usually carries the crop safely through the first severe frost, and gives them the advantage of two or three weeks good ripening weather, that usually precedes the severe frosts that usher in winter. I have known years when the maggots and bugs proved so destructive to the crop a few miles from the coast, as to bring squashes up to 40 and 50 dollars the ton, when at the sea side the crop was as large as usual, having received but little or no injury.

SQUASHES FOR STOCK.

When a large quantity of squashes is stored, there will always be more or less of waste. If in a large town, many of the spotted squashes can be most profitably handled by cutting out the decayed portion, and marketing the squash at a reduced price. It has been my practice for years to dispose of many of my defective squashes in this way, and I would state, as a very fair index of the comparative popularity of the Autumnal Marrow, Turban, and Hubbard squashes, in a community where they have all been grown for years, and are well known, that the sales of my market-man would average, late in the fall and in early winter, ten pounds of Hubbard and Turban to one pound of the Marrow, though he offered the Marrow at one-third the price of the Hubbard and Turban. After many trials I have found it next to impossible to dispose of the Marrow, while having a stock of Hubbard and Turban, hence have adopted the plan of feeding the former to my stock.

I have fed principally to horned cattle and pigs. The squashes should first have the seed removed, as these tend to dry up milch-cows, or, if fed to pigs, to cause them to urinate very freely. The Marrow should be fed to horned stock either in quite large pieces, or in pieces about three inches square, to prevent choking—for, if made much smaller, the cattle are more liable to choke. The Hubbard should always be cut into pieces three inches square, as the shell and curve of large pieces combined, are too much for the cattle to manage.

If squashes are plenty, they may be fed very liberally, a bushel and more a day for each head; the only danger to be guarded against being lost they relax the animals too much. In value for milk purposes, they appear to combine the good qualities of the Mangold Wurtzel, and the Carrot, both increasing the flow of milk and improving its quality. This is more particularly true of the Hubbard and Turban varieties. For fattening purposes, the Hubbard is excellent, as might be anticipated from the large proportion of sugar which is developed in it at the approach of winter. I have known a cow to be fatted for the butcher on the Hubbard squash, used in connection with good English hay.

In feeding to pigs, it can be fed.raw, or be boiled up with meal, or meal and scraps. My usual practice has been, to boil the squash in a Mott's boiler, about a barrel and a half at a time, adding a peck of beef or pork scraps, broken into small pieces, and stirring in meal, sufficient to thicken it. When cooked, it should be cooled as soon as possible, as the squash is very apt to sour, and make the mass thin and somewhat unpalatable to the animals. I have known a sow, with young, to be kept wholly on raw Hubbard squashes, and on her coming in to be in better condition than was desirable.

Squashes might be raised for cattle among corn as pumpkins are, (they are better food for animals than pumpkins,) though I have doubts of the profitableness of this double crop, where each makes its growth and matures at about the same time.

No doubt an improvement on this is, to omit every third row of corn, and give the vacant space to the squash hills. Among seed onions, I grow squashes with little or no apparent detriment to the crop, but in this case the crops are planted and mature with more than a month's difference between them at each end of the season. Besides horned cattle and hogs, many horses, goats, poultry, and rabbits will eat squashes with avidity.

As to their comparative value as food for stock, each

grower must strike the balance for himself—the facts being, that the yield is from one-fourth to one-third as great as carrots, and from one-fourth to one-fifth as great as mangolds, while they require but a fraction of the care in cultivation and gathering, that either of these crops do.

VARIETIES OF SQUASHES.

Owing to the great tendency in the varieties of the Cucurbitaceous Family to cross with each other, hybids are very common. Seed planted the first season after the crossing has been made, will usually produce a greater crop than either of the parent kinds, and individual squashes will be superior in quality to either of the parents; yet, as a rule, hybridization is not desirable, for, after the first season, there is a deterioration in the quality, below the average of the parent kinds, while the mixed varieties are not so marketable as the pure kinds.

Hubbard Squash.—I have traced the history of this squash back about sixty years, when the first specimen

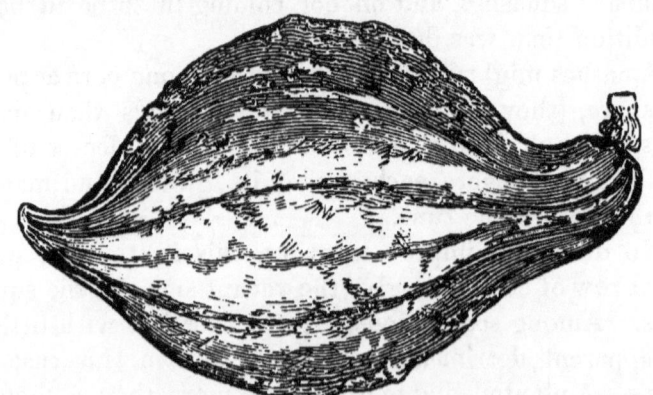

HUBBARD SQUASH.

was brought into Marblehead by a market-man named Green, who lived in the vicinity of Boston. The person who, when a girl, ate of the first specimen, is now living, an old lady of over four score years, and recalls the original form, which is much like that of the present type—turned up "like a Chinese shoe." It is now above twenty years since the variety was first brought to our notice by our old washerwoman named Hubbard; and to distinguish it from a blue variety that we were then raising, we called it "Ma'am Hubbard's Squash"; and when the seed became a commercial article, and it became necessary to give it a fixed name, I called it the Hubbard squash. If I had been able at the time to forecast its present fame, and have foreseen that it would become the established winter variety, throughout the squash growing region, I might have bestowed some more ambitious name; and again I might not, for the old lady was faithful in her narrow sphere in her day and generation, a good, humble soul, and it pleases me to think that the name of such an one has become, without any intent of hers, famous.

The form of the Hubbard is spherical at the middle, gradually receding to a neck at the stem end, and to a point usually curved at the calyx end, where it terminates in a kind of button or an acorn. In color it is dark green, excepting where it rests on the earth, where it is of an orange color. It usually has streaks of dirty white beginning at the calyx end, where the ribs meet, and extending half or two-thirds way up the squash. After the squash ripens, the surface exposed to the sun turns to a dirty brown color. The surface is often quite rough, and presents quite a knotty appearance. When the Hubbard is ripe it has a shell varying in thickness from that of a cent to that of a Spanish dollar.

For a year or two after we began to cultivate the Hubbard, we cultivated also a blue colored squash, called, at the time, the Middleton Blue. In a few years this squash became so thoroughly incorporated with the Hubbard, by repeated crossings, that it appeared to share the characteristics of a new variety; hence we called it the blue Hubbard, and for some years I spoke of two varieties of the Hubbard, a green and a blue kind. On testing the blue variety by itself, I found it had the characteristic of all hybrids, a tendency to sport. For this reason, of late years I have endeavored to throw it entirely out of cultivation in my seed stock.

After the Hubbard squash became somewhat noted, a communication occasionally appeared in the Press claiming that it was but an old variety revived. After giving all these claims, including those made to me personally by private correspondence, a fair examination, I am persuaded that the Hubbard is not an old variety revived, and that until it was sent out from Marblehead, with the exception of such cases as could be traced to seed distributed occasionally by me during the course of few years previous, it it was unkown in the United States. In my endeavors to trace its origin, the nearest I have come to it was in a variety of squash procured from one of the West India Islands, which had many characteristics in common with the Hubbard, though the shells of the squash were uniformly blue in color, and its quality was somewhat inferior.

Several claimed that it was but the Sweet Potato squash revived. I have raised a squash called by that name myself, and have seen two or more other lots that were raised by friends, from seed procured in different sections of the United States, and never saw one yet that resembled the green Hubbard.

The apparent connection between the Sweet Potato and Hubbard squash, I am convinced, has been made through the blue variety, which, when without a shell, has a close resemblance to some of those kinds that go under the name of "Sweet Potato" squash.

American Turban Squash.—I have given the prefix *American* Turban Squash, to distinguish it from the French Turban, with which many seedsmen have confounded it. The French Turban is the most beautiful in color, and the most worthless in quality of all the varieties of squash that have come to my notice. Nearly flat in shape, growing to weigh ten to twenty pounds, it has a large prominence at the calyx, and shaped like a flattened acorn; this is elegantly quartered, with a button in the middle, and is most beautifully striped with white and a bright grass green, while a setting of bead work surrounds it. The body of the squash is of the richest orange color. In quality the French Turban is coarse, watery, and insipid.

AMERICAN TURBAN SQUASH.

The *American* Turban is, without doubt, a combination of the Hubbard, Autumnal Marrow, Acorn, and French Turban, and the finest achievement that has as yet been obtained by hybridization. Like all hybrids it tends to sport, and varies somewhat in quality, so that while most of the squashes are of first quality, some will be found that are inferior; yet, with such parents as the Hubbard, Acorn, and the Autumnal Marrow (when we recall its early excellence), we might expect to find a superior squash, and in the *average* quality of the Turban we shall not be disappointed, for in dryness, fineness of grain, sweetness, delicacy of flavor, and richness of color, when fully ripened, it cannot be surpassed. Like the Hubbard, it is edible before it is fully ripe, either of these varieties, particularly the Hubbard, being superior for table use when unripe to any of the varieties of summer squashes. The form of the body of the squash is nearly cylindrical, the two diameters being usually in the proportion of three to five, while it is more or less flat at both the stem and calyx ends. At the calyx end there is usually more or less prominent an acorn. This may be very clearly defined, standing out very prominently from the body of the squash, or it may be very much flattened and sunk within the body, with the acorn barely traceable. In degree of prominence the acorn sports greatly, for on squashes growing on the same vine, I have found in one specimen the acorn projecting very prominently, and very fully developed, while on a second specimen it could only be traced in a rudimentary form. It is not desirable that the acorn should be prominent, as the seed extends into it at the calyx end of the squash where the meat is very thin, and if the acorn is very prominent, a slight bruise will injure it and cause the squash to rot. For this reason I have selected seed squashes for the last two or three years from those in which the acorn was not very prominently displayed, endeavoring to produce a type in which it should be little more than rudimentary.

Some writers on vegetables treat the American Turban squash as but an improved form of the French Turban, whereas it is a distinct variety. It is indebted to the French Turban for nothing more than the principal features of its form, getting its quality, keeping properties, color and fineness of grain from its other parent. As the American Turban is the result of hybridization, there is more or less of variety in the shape and color of the crop, and this will continue to be so unless by long and close cultivation of a particular type, sufficient individuality shall be acquired by this one type to stamp the entire crop. Though it may be a very pleasing thing to the eye to see every specimen alike, yet I consider it too great a risk to cultivate a hybrid squash for this end; for who knows what characteristics each parent has contributed or how much these are affected by each other in combination? Until these points are determined, there is danger, lest in continued selections of a given type some good traits should be eliminated.

We know that in some way the original excellence of

the Autumnal Marrow squash has been lost, and no one can, for a certainty, tell when or how this disappeared, and though originally an admixture of other sorts was doubtless the first step towards this deterioration, yet we are inclined to believe that a tendency to give prominency and individuality to the original admixture, has gradually borne under the good traits of the original Marrow.

Autumnal Marrow Squash.—This is also known as the Boston Marrow, or Marrow, it having been a very prominent squash in the markets of Boston for a series of years. A mongrel early variety of it is also known as the "Cambridge Marrow." This squash was introduced to the public by Mr. J. M. Ives, in the years 1831-2. When in-

AUTUMNAL MARROW SQUASH.

troduced, it was a small sized squash, weighing five or six pounds, fine grained and dry, with an excellent flavor. Marketmen found that by crossing with the African and South American varieties, they could increase the size of the original Marrow; they did this without troubling themselves about any risk of deteriorating the quality, and I doubt not that much of the present inferior quality of the Marrow squash is due to this vicious crossing. In form the Marrow is much like the Hubbard, but with less distinctive prominence in the neck and calyx. In color, the Marrow is between a lemon yellow and a rich orange;

the skin is covered with fine indentations, giving it a pock-marked appearance. The body of the squash is divided into sections by slight depressions in its longest diameter. Under the thin outer skin, or epidermis, is a thicker skin of a dark orange color. The flesh is orange colored. The seeds are somewhat larger and thicker than in the Hubbard, and considerably larger but not so thick as in the Turban. In quality the Marrow of to-day varies much; sometimes we find specimens that are all that can be desired, particularly as we get near to the original type, (this has been kept more nearly correct in Marblehead than elsewhere), but in its general character the Autumnal Marrow is watery, not sweet, and oftentimes deficient in flavor and fineness of texture. From its great productiveness, it is a favorite squash with marketmen, and its rich orange color and handsome form render it popular with those who have not become acquainted with the more recently introduced and finer varieties. There are two varieties grown extensively for Boston market known as the Cambridge Marrow. One of these is quite large in size, usually having the green color at the calyx, indicating a mongrel variety; the other is of medium size, and is characterized by a brilliant orange color, that makes it very attractive to the eye. Both of them mature a little earlier than the purer sort.

These three varieties of fleshy stemmed squashes, the Hubbard, American Turban, and Autumnal Marrow, include most of those raised for market purposes. There is a large number of other varieties, such as the Valparaiso, African, Honolulu, Cocoa-nut, Sweet Potato, etc., some of which have quite distinct characteristics, that are more or less raised in the family garden, but several of them are of inferior quality, some are hybrid, and though one or two may be desirable for the garden, yet none of them, as far as I have made acquaintance with them, have characteristics which would invite their general cultivation.

In that excellent work by my friend, Fearing Burr, "The Field and Garden Vegetables of America," will be found quite a list of summer, fall and winter varieties. I am often in receipt of varieties of high local repute in different sections of the country, and it is possible that some of them when tested may prove worthy of general cultivation.

Passing to the hard or woody stemmed varieties, we find included among them the Winter Crookneck, the Canada Crookneck, Yokohama, and Para.

The Crooknecks had their day and generation before the introduction of the soft-stemmed varieties. They were then the standard sorts, and the kitchens of thrifty farmers were adorned with choice specimens hanging suspended around the walls by strips of list, to be used during the winter, in the course of the spring, and even well into the summer months. The Crooknecks are characterized by long, usually curved necks, terminating in a bulb-like prominence at the calyx end, which contains the seed. The vines are covered with rough spines, and in the shortness of their leaf-stalks, the smaller size and different color of the leaves, are readily distinguished from the soft-stemmed sorts. They vary much in color at the time of the gathering, and there is a general tendency in all of them to change to a yellow hue in the course of the winter. In quality, the Large Winter Crookneck is coarse grained and watery, while the Canada Crookneck is finer grained, and at times quite dry and sweet. The Winter Crookneck weighs from ten to twenty-five pounds and upwards, and the true Canada Crookneck, which is rarely found pure, averages from four to six pounds. In keeping properties, the Crooknecks excel, frequently keeping in dry, warm apartments the year round. and, in a few instances, two years. When kept into the summer the seeds are at times found to have sprouted within the squash.

The Crooknecks are subject to a kind of dry rot, particularly in spring, which gives them a peculiar appearance when cut, the tissue between the cells having a dull, white color, though the fibres of flesh still retain their bright yellow color. Worthless for table use. The true measure of the length of time a squash keeps, is how long it keeps its *quality*, and not its mere structure.

CROOKNECK SQUASH.

The **Yokohama** is comparatively a new visitor from Japan, it having been received in this country in the year 1860, by Mr. James Hogg, from his brother then residing at Yokohama in Japan. The vine is a very free grower and a good yielder, though from the comparatively small size of the squash, the weight of the crop is not large when compared with the Hubbard, Turban, or Marrow. It is quite flat in shape, with somewhat of a depression at each end. The diameters are to each other about as one to three or four. It is deeply ribbed, and the flesh, which is of a lemon color, is remarkably thick, making it the heaviest of all squashes in proportion to its size. The flesh is very fine grained, smooth to the taste, and has a flavor resembling the Crookneck. With those who like the taste of the Crookneck, the Yokohama will likely be very popular.

In external color, before ripening, it is of an intensely dark green, covered with blisters, like a toad's back; as it ripens, it begins to turn of a light brown color at both the stem and blossom ends, and, after storing, it soon becomes entirely of a copper-like color, and is covered with a slight bloom. It may be well to start this squash under glass, on squares of turf, though, after an experience of three seasons, I am pursuaded that it is becoming acclimated; indeed, my crop of last season ripened with the

Hubbard and Turban. The cultivation of the Yokohama is mostly confined, as yet, to private gardens.

Para, or Polk Squash.—This is a half-bush squash. In the first stages of its growth, it has a bush habit, and sets its first fruit like a bush squash, but later it pushes out runners eight or ten feet in length, and bears fruit along them. The squash was brought to this country from Pàra, in South America. In shape it is oblong;

PARA, OR POLK SQUASH.

it is ribbed, of a tea-green color, excepting the portion which rests on the ground, which is of a rich orange color. The squashes weigh about three pounds each. They require the whole season to mature, and when in good condition, the flesh is dry and of a rich flavor. Like the Yokohama, I apprehend they will be very popular with a class, rather than with the community at large. Both the Yokohama and the Para can be kept well into the winter. I have kept a Yokohama, crossed on the Turban, fourteen months, and Hubbards, in two instances, twelve months.

THE SUMMER SQUASHES.

The remarks made relative to the cultivation of the fall and winter varieties, will apply to the cultivation of the summer squashes, with the exception of the distance between the hills; this, as they are of a bushy habit, should be about five feet. In quality, the summer squashes have but little to recommend them; it is principally their fresh, new taste that makes them acceptable for the table. South of New York, the cultivation of squashes is confined almost wholly to the bush varieties. Until recently, the New York market for fall and winter squashes has been supplied largely by the growers around Boston.

I find that there is a strong belief among prominent seedsmen in the Middle States, that the running varieties of squashes will not succeed in their section—they will not form the thick, fleshy root, they say. We, in the North, have always looked upon the squash as a half tropical fruit, and anticipated finding greater and greater success in its cultivation, the farther South it was planted. It has all the characteristics of a semi-tropical plant, like the tomato and melon, and should it be true that there is such a climacteric limitation, it would be a marked exception to a general law. I presume a canvass of my correspondence would settle the question, and regret that I have not time to do this; yet I have but little doubt that, under proper culture in the South, our running varieties would do as well, or better, than they do North. It occurs to me, at this moment, that Dr. Phillips, the enterprising editor of the Southern Farmer, stated to me, in the course of correspondence, that he had raised them by the acre in Mississippi with complete success.

The standard summer varieties are the Yellow and White Bush Scollop, often called Pattypan or Cymbals, and the Summer Crookneck. Of these the Summer Crookneck is the best. All of these form a shell as they ripen, and are then unfit for the table. They should not be cooked after the shell can be felt by the thumb-nail. The Green Striped Bergen is an early variety, quite popular in the markets of New York. A small squash, about twice the size of a large orange, some-

WHITE-BUSH SCOLLOPED SQUASH.

what fluted, called Sweet Potato Squash, is highly prized by some who are of high repute among squash fanciers. Several of the varieties that are grown as gourds, for ornamental purposes, are edible, a large proportion of them, indeed, as I have found on testing the largest of my specimens before feeding to the pigs. As a general rule, all that are not bitter to the taste are edible.

The Vegetable Marrow is about the only variety of the squash family cultivated by our English cousins. With them, it is brought to the table in the same style as our own varieties, or so cooked as to form part of a soup.

A friend, who resided some years in England, informed me that one of the greatest novelties to an English eye was an Autumnal Marrow Squash, which he kept as a center piece on his marble table for a month or more.

The Custard Squash, one of the hard stemmed sorts, of a yellowish cream color, oblong in shape, deeply ribbed, weighing from twelve to twenty pounds, is quite a favorite.

ENEMIES OF THE VINE.

The insect enemies are the striped bug (*Galeruca vittata*), or pumpkin bug (*Coreus tristis*), and the insect that produces the squash maggot. The striped bug appears about the first of June, and several broods being hatched in the course of the summer, they continue their depredations throughout the season. After the vines have pushed their runners two or three feet, their vigor is such that the after depredations of this little insect is of no practical importance—with the exception of injury occasionally done to immature squashes, the upper surface of which are sometimes found covered with them, and hundreds of little cell like holes are eaten out. The injury done by the striped bug is mostly confined to the period in the growth of the vine between its first appearance above the ground and the formation of the fifth leaf. They feed on both the upper and under surface of the leaf, and, sucking its juices, soon reduce it to a dry, dead net-work. The eating of the seed leaves of the plant, the two leaves which first appear, is not always fatal, provided the leaf that starts from between them is uninjured; if this, however, is eaten out, for all practical purposes the plant is destroyed, and should be pulled up and thrown away, no matter if the seed leaves are wholly uninjured. In those localities where the striped bug is not very prevalent, the greatest harm of its ravages is sometimes prevented by planting the seed about the tenth of May, should the weather permit, which will enable the vines to get so far along as usually to be beyond the reach of serious injury. The preventives to the ravages of this little insect, which attacks the whole vine family, including cucumbers and melons, are numerous. They may nearly all be brought under two classes: those which act mechanically, by covering the leaves so as to make them inaccessible to its punctures, and those which repel the insect by their disagreeable odors or pungent flavor. The best protectors of the first class are hand glasses, little frame-works covered with millinet or some very coarse cotton cloth, or, as this insect usually flies but a few inches above the surface of the earth, any box, circular or square, from which the bottom has been removed, having sides about ten inches in height. The remedies of the second class are those which are principally relied on where squashes are cultivated on a large scale. These should be applied early in the morning when the dew is on, or directly after a rain, when the leaves are wet, that they may adhere. In using them a small fine sieve will be found very convenient. The best of these remedies I name in the order of their popularity in great squash-growing districts. Ground plaster, oyster-shell lime, air slaked lime, ashes, soot, charcoal dust, and common dust. Plaster and oyster-shell lime I consider of equal value, and the use of protectors in my own grounds is confined to one or the other of these. Against air slaked lime, which is very commonly used, there is this serious objection.

However thoroughly it may be air-slaked, it still remains sufficiently caustic in its nature to seriously injure the leaves, causing more harm by its burning properties than good, by preventing the ravages of the bug. I have seen an acre of thrifty vines entirely destroyed, through the caustic properties developed in the lime by a gentle shower that fell just after its application; the leaves were so burned that they rubbed to dust in the finger. Charcoal dust and soot not only protect the vines, but serve also to draw the heat of the sun, oftentimes very grateful to the young vines in the early season of the year; while soot and ashes in all localities, and plaster and lime in some localities, as they are washed from the leaves by the rain, serve as a stimulating manure to the young plants. The advantages of plaster and oystershell lime are, that being very finely powdered, they can be easily dusted over the vine, while their white color has the advantage that it can be seen at a glance whether the leaves are fully covered. Common dust sounds cheap as a protector, but the trouble of collecting and separating from stones that might otherwise injure the leaves, is more than an offset to the cost of other articles. These protectors should be applied as soon as the young plant breaks ground, before it has fairly shaken off the shell of the seed, as the insect is often at work then, and the application should be renewed after every shower, the object being to keep every leaf entirely covered as far as practicable until the fifth leaf is developed, when the plants are usually beyond reach of injury from this little enemy, provided the hills have been supplied with rich, stimulating manure, sufficient to give them a rapid growth. Among this class of remedies, watering the plants with a decoction of tobacco, a little kerosene oil, stirred in water while being applied,

3*

(the proper proportion of this had better be tested by experiment), applying water in which hen manure or guano has been dissolved, sprinkling the leaves with a mixture of wheaten flour and red pepper, or snuff, or sulphur, etc., etc., have been found efficacious by various persons. Dr. Harris states that these insects fly by night as well as by day, and are attracted by the light of burning splinters of pine knots, or of staves of tar barrels. As insects breathe through pores in their bodies, such strong ammoniacal odors as are given off from a liquid in which hen manure, guano, or kerosene have been mixed, must tend to suffocate and so repel them.

As new land is much less infested with bugs than old land, in sections where these insects are very troublesome, it will be better to break up sward.

In fighting these pests, where but few hills are cultivated, pieces of board or shingle laid around the young plants, just above the surface of the ground, will collect many on their undersides over-night, and by examining them early in the morning, many can be brushed off into hot water. I don't think much of the plan of killing them about the vines; the old saying that "when one is killed fifty will come to his funeral" appears to have a savor of truth in it, for I have noted that where I have killed them about the vines, there seems to be no end to the business; with constant attention, still the bugs appear to be about as plenty as at first. I think that the odor from the dead ones attracts others.

The large black bug I consider rather a pumpkin than a squash bug, as in this section, and in others, as far as my knowledge extends, where the cultivation of the pumpkin has been given up for a number of years, it has almost entirely disappeared. Occasionally a leaf of a vine will be seen pretty well covered with the rascals late in the season, but so scarce are they that for several years past I have not seen, on an average, more than one a season on my vines, and I cultivate several acres annually. When the plants are young, they are likely to be found, if at all, below the elementary leaves, sucking out the juices from the vine itself. For these fellows there is nothing like finger work. I have known an instance in the interior where they were so numerous on Pumpkin vines planted among corn, that the mere smell of them acted as an emetic to three separate sets of hands that attempted to hoe the corn patch.

The squash maggot is hatched from the egg of an insect bearing a close resemblance to the lady-bug, but of a size considerably larger. The eggs are usually deposited near the root of the vine, within an inch or two of the ground; and in seasons when this insect abounds, eggs are deposited at the junction of the leaf stalks with the vine along some six or eight feet of vine. As soon as the egg is hatched, the maggot begins to eat his way through the center of the vine, and his boring will be seen outside his hole, like those of an apple-tree borer. The vines thus attacked will wither under a mid-day sun, and the injured ones are thus readily detected. Squashes on such vines usually make but little growth, and the vines ultimately die. If the presence of the borer is early detected, he can sometimes be killed by thrusting a wire, or stout straw into his hole; sometimes the vine is slit open and the intruder found and killed, but vines thus treated do not always recover. If the slit portion is covered with earth and pegged down, sometimes but little injury is done. I have taken thirteen borers from a single vine, some of the largest being an eighth of an inch in diameter and an inch in length.

It happens, at times, after the vines have made a vigorous growth of several feet, they suddenly wilt and die without any perceptible cause; no insects are to be found on the leaves, there are no borers in the vines, and on examining the roots, everything to be seen by the naked eye appears sound and healthy. I am at a loss to explain the cause of this, unless it be that the vine has been poisoned by something that it has taken into its circulation. I have picked half-grown plums from a tree that tasted as salt as brine. The tree had received a heavy manuring with salt, and ultimately died, proving that there is such a thing in the vegetable world as a tree poisoning itself by feeding to excess on one variety of food; and what is true of a tree may be true of a vine.

WOODCHUCKS AND MUSKRATS.

On low land, near water courses, Muskrats will sometimes make sad havoc with the growing fruit; while on uplands, the Woodchuck is sometimes exceedingly destructive. If the portion troubled by muskrats is of small area, the squashes can be protected by taking boxes of sufficient size, cutting a narrow slit in their sides, and setting the squashes in them, having the vines enter and go out of the narrow slits. When muskrats begin on a squash, as far as I have observed, they make a finish of it before injuring others.

Woodchucks are exceedingly destructive; they rarely entirely devour a squash, but gnaw more or less all in the vicinity of their burrows. If these burrows are not conveniently near the squash patch, they will leave the old and make new ones close by, or even in the midst of the squash field. The wounds made by their broad teeth soon heal, if the squashes have not reached their growth, and the gnawing has not been through the squash, but the crop is much injured for market purposes, and the squashes are apt to rot at the gnawed places after they are stored. I have had a ton injured in this way one season by a single woodchuck. A thousand-and-one ways are given to catch and destroy the woodchucks; traps set a little way down in their holes, and carefully hidden with earth, and apples containing arsenic, rolled into their burrows, are among those that have proved successful. It is worth while to offer five dollars for the skin of a woodchuck that has commenced depredations in a squash field.

SAVING SEED.

In selecting squashes for stock seed, take, while the squashes are in the field, or immediately after they are gathered, neither the largest nor the smallest specimens. The largest specimens are very tempting, particularly so if they have the true form, appear to be well ripened, and, if Hubbards, have a hard shell; but experience has proved that these, as a class, are most likely to be of impure blood. About a year ago two of my neighbors, who had become famous for their large Hubbard squashes, came to me to get a new stock of seed to start from; they stated that within a few years a large proportion of their squashes grew soft-shelled. Now, as they had made it a rule to select the largest specimens for seed, I have no doubt but that the admixture that was evident, from the loss of the hard shell characteristic of the true Hubbard, had crept in that way. Every old squash grower is aware of the great change that has come over the Autumnal Marrow squash. When introduced, it was of small size, weighing about five or six pounds, exceedingly dry, fine grained, and rich flavored. Now its quality is uncertain, for the most part greatly deteriorated below the original standard, but *it grows to double the average size* of the original squash. I have not the slightest doubt but this deterioration is due to the vicious practice of saving seed stock from the largest specimens grown, these specimens having got their extra size from larger and coarser varieties of the African or South American type. If any one has doubts of this theory, he can easily satisfy himself by examining the calyx end of a crop of the largest sized variety of Marrow squashes, when he will find a proportion of them with the green color stolen from the African or South American family.

Having decided on medium sized specimens for seed stock, select those that are most strongly marked externally with the characteristics of the variety. If a Hubbard, it should be very thick and hard shelled, of a dark green color, and the rougher and more nubbed the better. Let it have a good neck and calyx end, and be as heavy in proportion to its size as possible. The stem of both this and the Marrow squash should stand at quite an angle with the squash, and have a depression where it joins, as this indicates an early ripened specimen. The flesh should be hard, fine grained and thick, and not stringy on the inside. See to it that the squash swells out to a fair degree in the middle, and has an average proportion of seed. Having selected such specimens as these, bring them to the final test of the dinner table, and reject every one that does not there show all the characteristics of dryness, flavor, and fineness, that belong to a first-rate specimen.

I know that the injunction to select specimens that swell out to a fair degree in the middle, is contrary to the course pursued by most farmers; yet I advise it on the ground that such squashes, having a good quantity of seed, have superior vitality and individuality, and are nearer nature's ideal of perfection in the animal and vegetable kingdom, being better able to maintain the species.

I have seen the working of this law most conspicuously in the Crookneck family of squashes. The cultivator's type of a fine market squash is one with as large a neck and as small a seed end as possible. Following out this idea, they select for seed, specimens with a small seed end, and the result, as far as I have observed, has been that the squash, in the course of a few years, has deteriorated and become worthless.

When to Take Out the Seed.—We have advised that the specimens for seed purposes be selected early in the season, because later, particularly when they have been exposed to a high degree of heat, the color becomes so changed that the work of selection becomes far more difficult. The next question to discuss is, when shall we seed them? Contrary to the generally received opinion, *the seed is not ripe when the squash is*—in other words, after the squash has completed its growth, the vines dying naturally and the stem being dead and hardened, still the seeds are not fully matured till sometime after the squash is stored. The length of time will vary with the season, it being longer in a wet season and shorter in a dry one, the two extremes being from one to three months. If seeds are taken out as soon as the squash is gathered, though at the time they present a very plump appearance, yet if they are examined after they are dry, a large proportion will be found to be plump only on one side, most of them to be twisted, and not a few of them entirely wanting in meat. When seeding large lots for market, I have found the percentage of loss in the weight of the seed quite an important matter, it being as high as one-fifth. After the squash is gathered, the process of ripening the seed goes on until the entrails are absorbed, or eaten up by the seed, and the seed continue to increase in plumpness and weight until their entrails are so far consumed that only so much remains as is necessary to hold together the seed structure. This final ripeness is indicated by the seed compartments in the squash becoming distinct, and the attachments peeling off like the skin from an orange. If, when the squash is opened, the seed are embedded in a hard, dense mass of growth within, that does not readily separate from the squash, they will be twice as hard to clean, and will weigh full twenty per cent. short of the weight of well ripened seed when cleaned.

The seed is cleaned from the intestines by being either

squeezed out or washed out. If squeezed out, it will dry sooner, and when rubbed and winnowed when dry will have a more velvety look than when washed. Where a large quantity is to be handled, it is cleaned more quickly by washing than by rubbing, but it requires to be dried upon a comparatively clean surface; whereas rubbed seed can be dried upon any surface, no matter how dirty, as the refuse squash that remains adhering to it effectually protects it from all injury. Washed seed should not be spread over one deep, and squeezed seed not over one and a half deep; each should be stirred after the second day. If washed seed is stirred earlier, it is apt to be injured by the tearing of the epidermis, which, for the first day or two, adheres strongly to the surface it is spread on. The temperature for drying seed should not be over about one hundred degress, and better less than higher. *Never dry seed in an oven*, or very near a stove. The upper shelf of a kitchen closet, or a plate on the mantle piece, not too near the stove funnel, are each of them handy, though housewives will sometimes say they are not suitable places—if mice are apt to gnaw the seed in the closet, or children to see them on the mantle, for a certainty I will not dispute them. When the quantity to be cleaned is small, the sooner it is attended to, after the entrails have been removed from the squash, the brighter the seed will look; but if the quantity is large, by letting the mass stand one or two days, until fermentation begins and the entrails are partly decayed, the seed can be cleaned with far greater expedition. Much care and some experience is requisite to determine how far fermentation can be allowed to advance. As a general rule, if, on thrusting the hand into the middle of the mass, it feels milk warm, it should be at once mixed well together, and the whole be washed out within six hours. The great danger in permitting fermentation to advance too far is losing the white, ivory-like epidermis of the seed, thus destroying much of their beauty, and lowering their value for market purposes. In washing the seed, the water used may be made about milk warm, and so soon as they have been squeezed out of the entrails, skim them off the surface, dropping them into a sieve about as coarse as a common coal sieve; when this is nearly full, dash over them a couple of buckets of water, giving them immediately a quick shaking, which will tend to work out through the meshes fragments of the entrails that were taken out with them. If the hand is thrust into a mass of freshly washed seed, it will collect a good many pieces of the entrails. After pouring the water on the seed, incline the sieve at a sharp angle, in order to drain off the water. After they are well drained, pour them out on a large piece of soft cotton cloth, and rub and roll them well to absorb as much of the moisture as possible. Now spread as above directed. Two good hands, with seed in the right state, will sometimes wash out not far from one hundred pounds of seed in a day.

When are Squash Seed Sufficiently Dry?—It took me a couple of years to learn a very simple rule by which this can be infallibly determined; meanwhile I suffered a great deal of anxiety, took a great deal of extra care, (I got out twenty-six hundred pounds of squash seed one season,) and yet after all had a feeling of uncertainty in the premises. The ordinary way is to call squash seed dry when the enveloping skin has separated from the seed, and the seed itself is much contracted and has a dry look. If the temperature to which it has been exposed is quite low, this is a pretty safe guide, but if it has been dried at a somewhat high temperature—though the seeds may rustle with quite a dry sound when handled, yet appearance is a very deceitful guide—and if such seed are packed in barrels, they will be very likely to sweat, and when turned out, come out in caked masses, and if left together, will soon become musty. Squash seed, to be really dry, must be so in the meat as well as in the shell, and this can be in a moment determined by endeavoring to bend them. If they are pliable, they are not yet sufficiently dry; if they snap instead of bending, they can be safely stored for future use.

How long will Squash Seed keep their Vitality?—Squash seed, like all other seed, are best kept in a cool place, where the air is dry and the temperature is as even as possible. I have found that of the same lot of seed, those which were kept in an open bag did not retain their vitality as long by a year as those which were kept in the same bag, but put up in paper packages.

I have known squash seed to be fairly good at six years old, and again to be worthless when but three years old, and with no perceptible difference in the getting out and method of keeping of the two lots. I would lay down the rule to always test squash seed before planting, if it be over two years old. This can be easily done by putting a few in a cup, with water sufficient to swell them, covering them with some cotton wool, to prevent evaporation, and placing the cup where the heat is gentle, near the stove or on the upper shelf of a closet.

If the oil that enters into the composition of the meat of the squash seed has become rancid, the vegetative power of the seed is destroyed. This is easily determined by breaking the seed, when the meat will be of a dark color, and have a rancid taste. Under such circumstances, the shell of the spoiled seed will be usually darker colored than that of good seed. In a lot of seed saved at the same time, a portion will be spoiled, while the remainder will readily vegetate, and some that to the eye and taste appear to be perfectly sound, will prove to be utterly worthless. The cause of the difference in either case I do not know.

The proportion of seed and entrails of squashes to their entire weight is less than is generally supposed. By tests, applied towards the close of February, a few years ago, I found that the weight of seed and entrails to the entire squash, in the Turban, was as 65 to 1000; and, in the Hubbard, as 55 to 1000. At that date the entrails had less weight than they would have shown earlier in the season.

INSTINCTS AND HABITS OF SQUASH VINES.

It seems hardly fitting to close this treatise without alluding to something higher than the mere pecuniary or culinary value of the squash family. In common with all the vegetable world, it has instincts which are both curious and wonderful. How singular it is that roots have power to push through the soil directly to the spot where the best food is found, descending, if necessary, below the plane of growth, or ascending above it to the very surface and developing a perfect mist of rootlets to catch up the decaying particles found under a small heap of rubbish! Still more wonderful are some of the instincts of the vine itself. Each tendril stretches out to catch hold of, and fasten to something by which it can support the vine, and rarely, if ever, will it make the mistake of catching hold of any but the best supporter within reach. Yet more and higher even than this is the instinct they develop. They not only reach out for a support, and make selection of the object to which to cling, but they will vary the direction of their growth through quite a number of degrees in pursuit of the particular object they have selected. To see this wonderful phenomenon in its most striking aspect, select a vine of some one of the mammoth varieties, under circumstances in which its most vigorous growth will be developed. Let every stick, weed, or the like, be removed from the vicinity of the main runner, and then thrust firmly into the ground a slip of shingle, not over half an inch wide, on one side of the vine, a few inches beyond the outstretched tendril that is always found near the extremity, noting with care at the same time the direction in which the extremity of the vine points. Within twenty-four hours it will be found that the vine has turned from its former course, towards the side on which the shingle is placed, while the tendril has turned towards the shingle and perhaps found and grasped it! In proof that this is no mere chance event, let the slip of shingle be now removed, and placed in the same relation to the vine as before, but on the opposite side. Within twenty-four hours the vine will be found to have turned from its former course and to be inclined towards the side on which the shingle is placed, while the tendril on that side has shown a corresponding instinct. Then study the tendril. It is most admirably adapted for its office; it is usually a compound spiral, one-half of it winding to the right and the other half of it to the left, thus combining the greatest strength with the greatest possible elasticity. As another illustration of its wonderful instincts, I have seen a squash vine run about ten feet along the surface of the ground, keeping its extremity within a few inches of the surface, until it passed under the projecting limb of a pear tree, which was about four feet above the surface of the earth; here it stretched up almost vertically towards the tree, until it had almost reached it, when, not having sufficient stamina to support it to a further effort, it fell over towards the ground, forming an arch. It immediately turned up with a second effort to reach the tree, made a second failure and formed a second arch, and with still another failure a third arch, by which time the extremity had passed out from under the tree, *when it kept on its horizontal growth the same as before it had reached the tree!* Such instincts are wonderful. How did the vine know the tree was above it, or that the slip of shingle was at either the right or left of it?

During the best growing weather the growth of the vine is very rapid. I have found, by actual measurement, that a vine of the mammoth variety grew above fourteen inches in twenty-four hours. Sometimes, during a season of drouth, a surprising tenacity of life is displayed. I well remember one piece of vines growing on a shallow spot above a ledge, where, during a season of severe drouth, I could find nothing but earth as dry as dust, close down to the ledge; yet these vines, for more than a week, would wilt and apparently dry up each day, to renew themselves with the dews over night. I have very rarely (and I have often examined them for this,) found the tendrils of the squash vine seizing on the Apple of Peru, (Stramonium,) a large weed quite common near the sea shore, of disagreeable odor and poisonous in its nature, when taken internally. Now, the Apple of Peru is very common in our squash fields, and presents the most stable support of all the weeds of the field. Then why this apparent antipathy?

I have endeavored to make my little treatise as complete a manual as possible. If, from the directions given, so delicious a vegetable as the squash shall be more generally and more successfully cultivated, I shall be well pleased.

THE
SMALL FRUIT CULTURIST.

BY

ANDREW S. FULLER.

Beautifully Illustrated.

We have heretofore had no work especially devoted to small fruits, and certainly no treatises anywhere that give the information contained in this. It is to the advantage of special works that the author can say all that he has to say on any subject, and not be restricted as to space, as he must be in those works that cover the culture of all fruits—great and small.

This book covers the whole ground of Propagating Small Fruits, their Culture, Varieties, Packing for Market, etc. While very full on the other fruits, the Currants and Raspberries have been more carefully elaborated than ever before, and in this important part of his book, the author has had the invaluable counsel of Charles Downing. The chapter on gathering and packing the fruit is a valuable one, and in it are figured all the baskets and boxes now in common use. The book is very finely and thoroughly illustrated, and makes an admirable companion to the Grape Culturist, by the same author.

CONTENTS:

Chap. I. Barberry.
Chap. II. Strawberry.
Chap. III. Raspberry.
Chap. IV. Blackberry.
Chap. V. Dwarf Cherry.
Chap. VI. Currant.
Chap. VII. Gooseberry.
Chap. VIII. Cornelian Cherry.
Chap. IX. Cranberry.
Chap. X. Huckleberry.
Chap. XI. Sheperdia.
Chap. XII. Preparation for Gathering Fruit.

Sent post-paid. Price $1.50.

ORANGE JUDD & CO., 41 PARK ROW.

AMERICAN POMOLOGY.

APPLES.

By Doct. JOHN A. WARDER,
PRESIDENT OHIO POMOLOGICAL SOCIETY; VICE-PRESIDENT AMERICAN POMOLOGICAL SOCIETY.

293 ILLUSTRATIONS.

This volume has about 750 pages, the first 375 of which are devoted to the discussion of the general subjects of propagation, nursery culture, selection and planting, cultivation of orchards, care of fruit, insects, and the like; the remainder is occupied with descriptions of apples. With the richness of material at hand, the trouble was to decide what to leave out. It will be found that while the old and standard varieties are not neglected, the new and promising sorts, especially those of the South and West, have prominence. A list of selections for different localities by eminent orchardists is a valuable portion of the volume, while the Analytical Index or *Catalogue Raisonné*, as the French would say, is the most extended American fruit list ever published, and gives evidence of a fearful amount of labor.

CONTENTS.

Chapter I.—INTRODUCTORY.
Chapter II.—HISTORY OF THE APPLE.
Chapter III.—PROPAGATION.
    Buds and Cuttings—Grafting—Budding—The Nursery.
Chapter IV.—DWARFING.
Chapter V.—DISEASES.
Chapter VI.—THE SITE FOR AN ORCHARD.
Chapter VII.—PREPARATION OF SOIL FOR AN ORCHARD.
Chapter VIII.—SELECTION AND PLANTING.
Chapter IX.—CULTURE, Etc.
Chapter X.—PHILOSOPHY OF PRUNING.
Chapter XI.—THINNING.
Chapter XII.—RIPENING AND PRESERVING FRUITS.
Chapter XIII and XIV.—INSECTS.
Chapter XV.—CHARACTERS OF FRUITS AND THEIR VALUE—TERMS USED.
Chapter XVI.—CLASSIFICATION.
    Necessity for—Basis of—Characters—Shape—Its Regularity—Flavor—Color—Their several Values, etc., Description of Apples.
Chapter XVII.—FRUIT LISTS—CATALOGUE AND INDEX OF FRUITS.

Sent Post-Paid.   Price $3.00.

ORANGE JUDD & CO., 41 PARK ROW, NEW-YORK.

GARDENING FOR PROFIT,

In the Market and Family Garden.

By Peter Henderson.

FINELY ILLUSTRATED.

This is the first work on Market Gardening ever published in this country. Its author is well known as a market gardener of eighteen years' successful experience. In this work he has recorded this experience, and given, without reservation, the methods necessary to the profitable culture of the commercial or

MARKET GARDEN.

It is a work for which there has long been a demand, and one which will commend itself, not only to those who grow vegetables for sale, but to the cultivator of the

FAMILY GARDEN,

to whom it presents methods quite different from the old ones generally practiced. It is an ORIGINAL AND PURELY AMERICAN work, and not made up, as books on gardening too often are, by quotations from foreign authors.

Every thing is made perfectly plain, and the subject treated in all its details, from the selection of the soil to preparing the products for market.

CONTENTS.

Men fitted for the Business of Gardening.
The Amount of Capital Required, and Working Force per Acre.
Profits of Market Gardening.
Location, Situation, and Laying Out.
Soils, Drainage, and Preparation.
Manures, Implements.
Uses and Management of Cold Frames.
Formation and Management of Hot-beds.
Forcing Pits or Green-houses.
Seeds and Seed Raising.
How, When, and Where to Sow Seeds.
Transplanting, Insects.
Packing of Vegetables for Shipping.
Preservation of Vegetables in Winter.
Vegetables, their Varieties and Cultivation.

In the last chapter, the most valuable kinds are described, **and** the culture proper to each is given in detail.

Sent post-paid, price $1.50.

ORANGE JUDD & CO., 41 Park Row, New-York.

www.ingramcontent.com/pod-product-compliance
Lightning Source LLC
Chambersburg PA
CBHW031727230426
43669CB00007B/279